with Probability and S

Part II

Student Text

First-Year Algebra via Applications Development Project

Zalman Usiskin, Director

Supported by
National Science Foundation Grant No. SED74-18948

Published by the National Council of Teachers of Mathematics
1906 Association Drive, Reston, Virginia 22091
1979

Comments or questions about this work are invited from any
and all interested parties. Address all correspondence to
Zalman Usiskin, Director, First-Year Algebra via Applications
Development Project, Department of Education, University of
Chicago, 5835 S. Kimbark Avenue, Chicago, IL   60637.

Library of Congress Cataloging in Publication Data:

First-Year Algebra via Applications Development Project.
   Algebra through applications, with probability and
statistics.

   Includes indexes.
   1. Algebra. 2. Probabilities. 3. Statistics.
I. National Council of Teachers of Mathematics.
II. Title.
QA152.2.F49   1979   512.9   78-23568
ISBN 0-87353-134-5 (set)
     0-87353-135-3 (part I)
     0-87353-136-1 (part II)

TABLE OF CONTENTS

CHAPTER 16: FUNCTIONS

REFERENCE INFORMATION

CHAPTER 9

SLOPES AND LINES

Lesson 1

Types of Graphs

Situation 1: A diet

Horace Heavy went on a diet. To check how he was doing, he graphed his weight at 8 AM each day. (You are reading the graph correctly if you notice that he weighed 86 kg on the 4th day of the diet.)

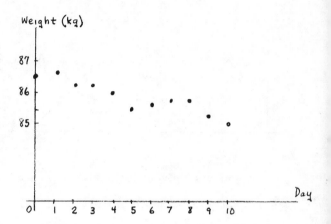

This graph is made up of points which are not connected. Such a graph is called <u>discrete</u>. But Horace has a weight throughout each day. His actual weight might lead to the following graph.

If you can draw the graph without taking your pencil off of the paper, the graph is called <u>continuous</u>.

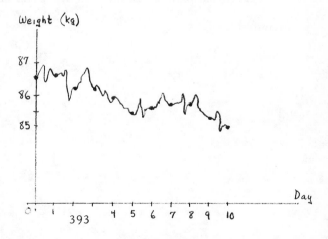

The previous graphs show that Horace's diet is working. If Horace's diet did not work, the graphs would look different. Here are three other possibilities.

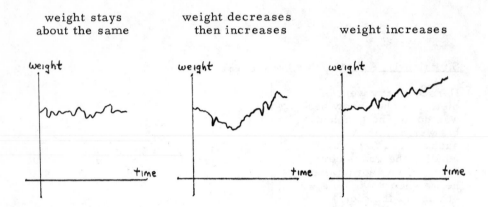

<table>
<tr><td>weight stays
about the same</td><td>weight decreases
then increases</td><td>weight increases</td></tr>
</table>

If you were given Horace's weight each day in a number, you could easily tell if his diet were working. But you could not easily see how much the weight changed, how big the changes were, and so on. Graphs are helpful because they enable you to use your eyes as well as your knowledge of arithmetic.

The above graphs are complicated. Some situations lead to smoother graphs.

Situation 2: Going 100 miles, how fast and how long?

The chart below gives some possible driving speeds. Across from the speed is the time it would take at that speed. The pairs are graphed and connected.

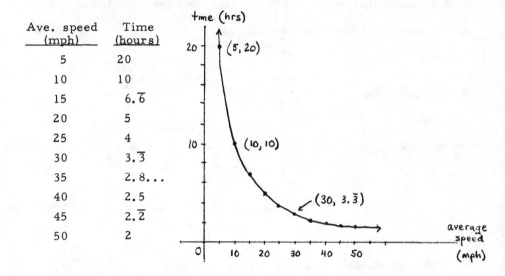

Ave. speed (mph)	Time (hours)
5	20
10	10
15	6.$\overline{6}$
20	5
25	4
30	3.$\overline{3}$
35	2.8...
40	2.5
45	2.$\overline{2}$
50	2

Look carefully at the above graph. It is steep until about 15 mph. This shows that at slower speeds, going 5 mph faster will save a lot of time. Traffic jams can really slow you up. At faster speeds, the graph is more level. At fast speeds, an extra 5 mph does not make that much difference in time.

<u>Situation 3: Mailing letters.</u>

The cost to mail a letter first-class depends on the letter's weight. Here is a graph which shows the 1976 costs. (The graph continues like this until 13 on the x-axis. No parcel weighing over 13 oz. can be mailed first class.)

This graph is not

discrete or

continuous.

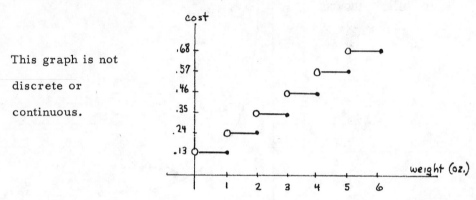

This graph shows many things about first-class mail. For example:

(a) Anything weighing less than or equal to 1 oz. costs 13¢ to mail.

(b) Anything weighing between 1 oz. and 2 oz. (including 2 oz.) costs 24¢ to mail, etc. (Each ounce adds 11¢ in cost.)

(c) It can not cost 18¢ to mail a first-class letter.

Questions covering the reading

1-3. Use these graphs.

(A)   (B)   (C)

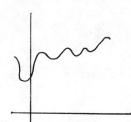  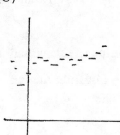

1. Which graph is discrete?   2. Which graph is continuous?

3. Which graph is neither discrete nor continuous?

4-6. Refer to the discrete graph of Horace Heavy (HH) on p. 393.

4. Approximately what was HH's weight at 8AM on the 9th day of the diet?

5. On which days did HH's weight increase?

6. Which day did his weight decrease the most?

7-9. Refer to the average speed-time graph on p. 395.

7. As you go to the right, from 0 to 15, the graph goes down steeply. What does this mean?

8. As you go to the right from 40 to 50, the graph is almost level. What does this mean?

9. According to the graph, what average speed will take you 12 hours to go 100 miles? (You will need to approximate.)

10. Make a graph of the average speed and time to go 200 miles.

11-14. Refer to the mailing cost graph in this lesson. According to the graph, what is the cost for first-class mail weighing:

11. $2\frac{1}{2}$ oz.   12. 3.01 oz.   13. 3 oz.   14. 4.63 oz.

15. Air mail letters to Mexico cost 17¢ for the first ounce, and 15¢ for each additional ounce. Make a graph which shows how the cost to mail these letters depends on the weight.

16. Why are graphs helpful?

Questions testing understanding of the reading

1-3. Multiple choice. Make your choice from these graphs.

(A)   (B)   (C)

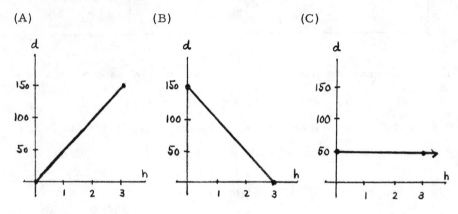

1. Which graph shows the distance d travelled in h hours at 50 mph?

2. You are 150 miles from home and travel home at 50 mph. Which graph gives your distance d from home after h hours of travel?

3. You are staying 50 miles from home. Which graph shows this?

4. Draw a graph which shows the distance d travelled if you travel for h hours at an average speed of 80 km per hour.

5. <u>Multiple choice</u>. If you threw a ball into the air, which graph
do you think best describes the path of the ball?

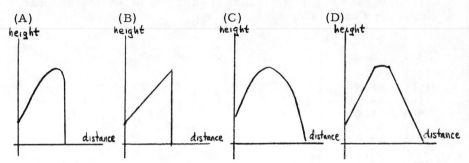

6. Refer to Question 5. Why don't the graphs begin at $(0, 0)$?

7-9. Suppose you have $50 and records cost $5 apiece. Use these
graphs.

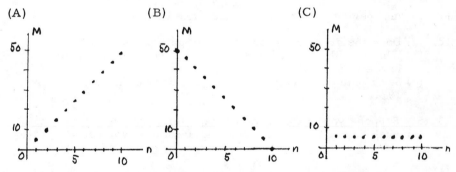

7. Which graph pictures the amount M you will have left after
buying n records?

8. Which graph pictures the total amount M you have paid?

9. Which graph pictures the cost M of the nth record?

10-12. Graphed here is the total value of gold (in millions of dollars) produced in the U.S. for some of the years from 1959 to 1970. (Value calculated at $35 per troy ounce of gold.) Source: International Monetary Fund, 1972.

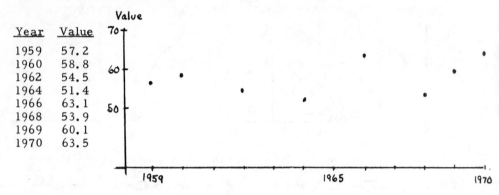

Year	Value
1959	57.2
1960	58.8
1962	54.5
1964	51.4
1966	63.1
1968	53.9
1969	60.1
1970	63.5

10. In what period did production increase the most?

11. In what period did production decrease the most?

12. (a) Does the graph show any trend in gold production?
    (b) What would you predict for the total value of gold produced in the U.S. in 1971?
    (c) See if you can locate an almanac which has the figures for 1971.

13-15. Suppose you graphed the average temperature where you live for each day of the year. What pattern would you get if you lived:

(A)   (B)   (C)

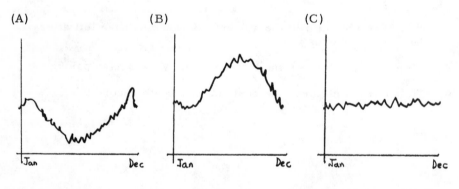

13. in Hawaii.   14. in Montreal.   15. in Buenos Aires.

16-20. Tell whether the graphs are continuous or discrete.

16. those in Questions 1-3

17. those in Question 5

18. those in Questions 7-9

19. those in Questions 10-12

20. those in Questions 13-15

Question for discussion

1. Graphs showing the change of a quantity over time (like Horace
   Heavy's weight) can often be found in newspapers. What quan-
   tities are often graphed in this way?

Lesson 2

Equations for Graphs

If a sentence contains one variable, each solution is a number.

Sentence: $3x < 15$ one solution: 2

If a sentence contains <u>two</u> variables, each solution is an <u>ordered pair</u> of numbers.

Sentence: $C = \frac{5}{9}(F - 32)$ one solution: $(32, 0)$

This sentence is a formula relating Fahrenheit and Celsius temperatures. We have picked F to be the first coordinate. C is the second coordinate. When F is 32, C is 0. The result is the ordered pair $(32, 0)$. For each Fahrenheit temperature, you can calculate the corresponding Celsius temperature.

F	C	Ordered Pair (F, C)
32°	0°	$(32, 0)$
0°	$-\frac{160^{\circ}}{9}$	$(0, -\frac{160}{9})$
68°	20°	$(68, 20)$
-40°	-40°	$(-40, -40)$
95°	35°	$(95, 35)$

The ordered pairs are graphed below.

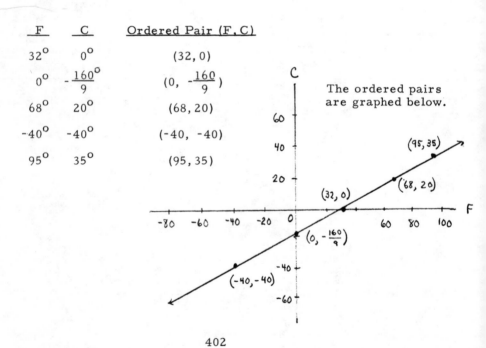

402

The points lie on a line. The line is <u>the graph of the solution set</u> <u>of the equation</u> $C = \frac{5}{9}(F - 32)$. For short, we call the line the "graph of the equation $C = \frac{5}{9}(F - 32)$."

A situation from Chapter 2 also leads to a graph where points lie on a line. Suppose some shirts sell for $6.50 each. Letting C stand for the cost of n shirts, $C = 6.50n$. Some of the many solutions are graphed below.

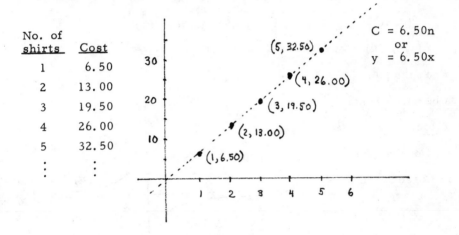

No. of shirts	Cost
1	6.50
2	13.00
3	19.50
4	26.00
5	32.50
⋮	⋮

Instead of writing $C = 6.50n$, it is common to write

$$y = 6.50x$$

Here x stands for the x-coordinate; y for the y-coordinate. Because each shirt is the same price and because the scales on the axes are even, the points lie on a line.

Not all equations lead to graphs which are lines. You already know how distance, rate, and time are related.

$$\text{rate (in mph)} = \frac{\text{distance (in miles)}}{\text{time (in hours)}}$$

If the distance is fixed at 100 miles, then

$$\text{rate} = \frac{100}{\text{time}}$$

Now let x stand for time, y for rate. Then

$$y = \frac{100}{x}$$

It is easy to find pairs of numbers which work.

$(10, 10)$ $(25, 4)$ $(4, 25)$ $(6, 16\frac{2}{3})$

$(-10, -10)$ $(-25, -4)$ $(-4, -25)$ $(-6, -16\frac{2}{3})$ etc.

When the points are graphed,

a beautiful 2-part curve results.

The curve is called a <u>hyperbola.</u>

Half of the hyperbola was graphed

in Lesson 1 on p. 395.

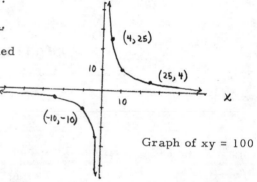

Graph of xy = 100

<u>Questions covering the reading</u>

1-3. Refer to the shirt-cost graph on p. 403.

1. Six shirts cost $39.00. This information gives what point on the graph?

2. Name 3 points on the graph of $y = 6.50x$ which are not pictured.

3. In the sentence $y = 6.50x$, x stands for which coordinate?

4-7. Refer to the F-C graph on p. 402.

4. (68, 20) is on this graph. What does that mean?

5. $10^\circ C = 50^\circ F$. What point on the graph is found by this information?

6. The points on this graph all lie on a line. What is an equation for that line?

7. Name 2 points which are on the graph but which are not pictured.

8-10. Consider the equation $y = 100/x$.

8. Name 5 points with integer coordinates which work in this equation.

9. Name 1 point with a non-integer coordinate which works in this equation.

10. The graph of _all_ points which work is called a(n) _____.

11. If a sentence contains two variables, then each solution is a(n) _____.

Questions testing understanding of the reading

1-6. Find and graph 5 solutions to each sentence.

1. $y = 10x$ 2. $y = 3x + 1$ 3. $x + y = 6$

4. $2x - y = 8$ 5. $xy = 4$ 6. $3x + y = 2\frac{1}{2}$

7. Let p be the perimeter of a square with side s. Then

$$p = 4s$$

Let s be the first coordinate, p the second. Graph 5 solutions to this equation.

8. Graph 5 solutions to the phone rate equation $C = .65 + .20(n - 3)$. Use n as the first coordinate, and only use integer values of n larger than 3.

Questions for exploration

1-4. Try to find an equation whose graph contains all of the given points. (You are not expected to have any regular way of doing this.)

1.

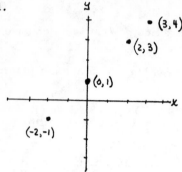

2.

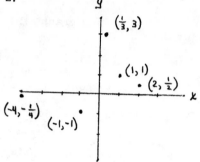

3.

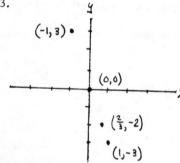

4.

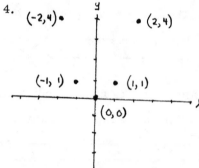

Lesson 3

Rates of Change

At age 9 Karen was 4'3" tall.
At age 11 she was 4'9" tall.
At age 14 she was 5'4" tall.

Did she grow faster from 9 to
11, or faster from 11 to 14?

To answer this question, we calculate <u>rates of change</u> of Karen's

height per year.

Rate of change in height per year (from age 9 to age 11)

$$= \frac{\text{change in height}}{\text{change in age}} = \frac{4'9'' - 4'3''}{(11 - 9) \text{ yrs}} = \frac{6''}{2 \text{ yrs}}$$

= 3 inches per year

Rate of change in height per year (from age 11 to age 14)

$$= \frac{\text{change in height}}{\text{change in age}} = \frac{5'4'' - 4'9''}{(14 - 11) \text{ yrs}} = \frac{7''}{3 \text{ yrs}}$$

= 2.$\overline{3}$ inches per year

So Karen grew at a faster rate from age 9 to age 11.

The data can be graphed and
points connected. The rate of
change then measures how fast
the graph goes up as you go to the
right.

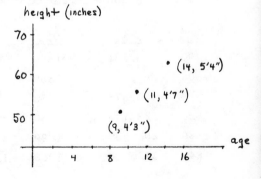

The segment connecting
(9, 4'3") to (11, 4'7") is steeper.
So there is a greater rate of change.

407

Past the age of 50, people tend to <u>lose</u> height as they get older. Suppose a man was 180 cm tall at the age of 55 and is 177 cm tall at the age of 65. He will have a negative rate of change.

His change in height per year

$$= \frac{\text{change in height}}{\text{change in age}} = \frac{177 - 180}{65 - 55}$$

$$= -\frac{3}{10}$$

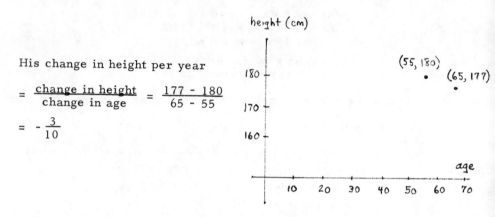

Every rate of change is found by dividing two changes. (Rate is one model for division.) The changes are directed distances and found by subtraction.

Here is the population of Manhattan Island (part of New York City) according to the ten-year census figures.

Year	Pop.	Year	Pop.	Year	Pop.
1790	33,131	1850	515,547	1910	2,331,542
1800	60,515	1860	813,669	1920	2,284,103
1810	96,373	1870	942,292	1930	1,867,312
1820	123,706	1880	1,164,673	1940	1,889,924
1830	202,589	1890	1,441,216	1950	1,960,101
1840	312,710	1900	1,850,093	1960	1,698,281
				1970	1,539,233

There is a lot of data. So a graph can help.

408

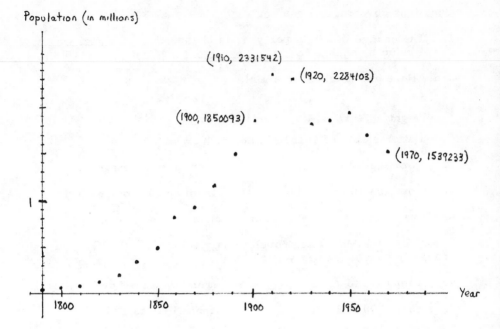

Population (in millions)

(1910, 2331542)

(1920, 2284103)

(1900, 1850093)

(1970, 1539233)

1

1800   1850   1900   1950   Year

A segment connecting the dots from 1900 to 1910 would be the most slanted upward to the right. That is when the largest rate of change of population occurred. That rate is

$$\frac{\text{change in pop.}}{\text{change in time}} = \frac{2,331,542 - 1,850,093}{1910 - 1900} = \frac{481,449}{10} = \frac{48,144.9}{\text{(per year)}}$$

From 1920 to 1970, Manhattan lost population. So the rate of change of population is negative.

$$\frac{\text{change in pop.}}{\text{change in time}} = \frac{1,539,233 - 2,284,103}{1970 - 1920} = \frac{-744,870}{50} = \frac{-14,897.4}{\text{(people per year)}}$$

Thus Manhattan lost an average of about 15,000 people per year in the 50 years from 1920 to 1970.

409

1. The change in height per year is change in _____ divided by change in _____ .

2-5. Here are heights for a girl.

age	9	10	11	12	13	14	15
height	4'3"	4'5"	4'10"	5'1"	5'3"	5'4"	5'5"

Calculate the rate of change in height (per year):

2. from age 9 to age 12   3. from age 10 to age 14

4. from age 10 to age 11   5. from age 12 to age 15

6. Give an example where a rate of change is negative.

7-10. Use the Manhattan populations given on p. 408. Calculate the change in population per year:

7. from 1840 to 1940   8. from 1910 to 1960

9. from 1790 to 1890   10. from 1920 to 1930

11-14. Use the information, p. 408, or the graph, p. 409.

11. When did Manhattan's population have the greatest increase?

12. When did Manhattan's population decrease the most?

13. Do you think Manhattan ever grew by 50,000 people in one year?

14. On the average, did Manhattan lose people at a greater rate from 1960 to 1970 or from 1920 to 1970?

Questions testing understanding of the reading

1. According to the Guiness Book of World Records, the most freakish rise in temperature ever recorded occurred in Spearfish, South Dakota, January 22, 1943. At 7:30 AM it was -4°F; at 7:32 AM it was 45°F. What was the rate of change in temperature (per minute)?

2. In Browning, Montana, January 23-24, 1916, the temperature fell from 44°F to -56°F in 24 hours. What was the rate of change in temperature per hour?

3. A road is 60 m above sea level at one point and 90 m above sea level at a second point 2 km away. Calculate the rate of change of height in this road from the one point to the other.

4. Six years ago there were 1250 students in the school. Eleven years ago there were 800. Calculate the rate of change in school population per year from eleven years ago to six years ago.

5. Accurately graph the data given in Questions 2-5 on page 410. Use the graph to answer the questions.

   (a) In which 2-year period did the girl grow the fastest? What was her rate of growth per year during that period?

   (b) In which 3-year period did the girl grow the fastest? What was her rate of growth per year during that period?

Lesson 4

Slope

A special type of rate of change is called <u>slope</u>.

Definition:

> The <u>slope</u> determined by two points (x_1, y_1) and (x_2, y_2) is the change in the y-coordinates divided by the change in the x-coordinates. That is,
>
> $$slope = \frac{y_2 - y_1}{x_2 - x_1}$$

All through Lesson 3 you were calculating slopes. Here are some examples which are repeated.

1. <u>Manhattan population from 1920 to 1970</u>: Think of the points as (1920, 2284103) and (1970, 1539233). Then x_1 = 1920, x_2 = 1970, y_1 = 2284103 and y_2 = 1539233. Using the above formula:

    $$slope = \frac{y_2 - y_1}{x_2 - x_1} = \frac{1539233 - 2284103}{1970 - 1920} = -14,897.4$$

    The slope is the rate of change of population per year. Since the population went down, the slope is negative.

2. <u>Rate of change in height</u>: Karen was 4'7" tall at age 11, 5'4" at age 14. The ordered pairs (graphed on page 407) are (11, 4'9") and (14, 5'4"). The slope formula gives the rate of change in

412

height per year.

$$\text{slope} = \frac{y_2 - y_1}{x_2 - x_1} = \frac{5'4'' - 4'9''}{(14 - 11)\text{yr}} = \frac{7''}{3\text{ yr}} = -2.3'' \text{ per year}$$

The name "slope" comes from the idea of the slope of a hill.

Diagrammed at right is one of San Francisco's steeper streets. For each unit across, the road rises .27 unit.

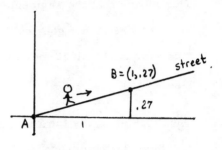

The slope determined by A and B is $\frac{.27 - 0}{1 - 0}$ or .27.

One General Motors test track for bulldozers goes down .60 unit for each unit across. The slope determined by C and D is $\frac{0 - .60}{1 - 0}$ or -.60.

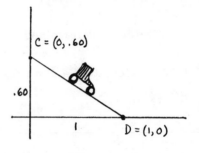

In these examples, the person and bulldozer travel one unit to the right. The slope is the amount of change in the height of the graph as you go 1 unit to the right.

The slope determined by (-5, 4) and (1, 3) is

$$\frac{3 - 4}{1 - -5} \quad \text{or} \quad -\frac{1}{6}$$

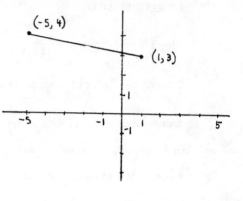

This means that the segment joining these points goes down $\frac{1}{6}$ unit for each unit it goes to the right.

Questions covering the reading

1. Slope is the change in _____ divided by the change in _____.

2. The slope determined by (x_1, y_1) and (x_2, y_2) can be calculated by the formula _____.

3-10. Calculate the slope for each pair of points.

3. (3, 6), (4, 10)   4. (5, 100), (7, 150)

5. (-1, 8), (2, -6)   6. (7, -5), (11, -2)

7. (1.2, -4), (-1.2, -5)   8. (8, -6.3), (1.4, -9.6)

9. $(-\frac{1}{5}, \frac{2}{3})$, $(-\frac{2}{5}, \frac{4}{3})$ 10. $(-\frac{3}{7}, -\frac{1}{5})$, $(-\frac{2}{7}, \frac{1}{2})$

11. Slope is a special type of rate of _____.

12-13. Calculate the slope for each pair of points. Tell what the slope stands for.

12. (12 yrs, 5'3"), (16 yrs, 5'10")

13. U.S. populations: (1960, 180 million), (1970, 205 million)

14-17. What would be the slope in each case?

14. A segment goes up 3 units for every unit it goes across.

414

15. A segment goes down $\frac{1}{2}$ unit for every unit it goes across.

16. A segment is horizontal.

17. A segment goes up 400 units for every unit it goes across.

Questions testing understanding of the reading

1. Multiple choice. The bulldozer test track in this lesson had a slope of -.60. Another way of saying this is:

    (a) The track has a downward slope of 6%.

    (b) The track has a downward slope of 16.$\overline{6}$%.

    (c) The track has a downward slope of 60%.

    (d) None of these.

2-3. Calculate the slope for each pair of points. Tell what the slope stands for.

2. Long distance: (5 minutes, $1.15), (10 minutes, $2.15)

3. High-rise apartment monthly rents: (13th floor, $325), (19th floor, $385)

4-5. Multiple choice. Use graphs (A), (B), and (C).

(A)   (B)   (C)

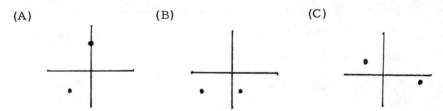

4. Which pair of points would lead to a negative slope?

5. Which pair of points would lead to a positive slope?

6. The slope determined by (200, 11) and (201, y) is 15. Find y.

7. The slope determined by (-6, 5) and (x, 2) is $\frac{1}{3}$. Find x.

8. True or False: The slope determined by (1, 5) and (11, -83) is the same as the slope determined by (11, -83) and (1, 5).

9. True or False: The slope determined by (2, 3) and (4, 7) is the same as the slope determined by (4, 7) and (2, 3).

10. True or False: $\dfrac{y_2 - y_1}{x_2 - x_1} = \dfrac{y_1 - y_2}{x_1 - x_2}$

11. True or False: You can switch the order of points without changing the slope that you calculate.

Questions for discussion

1-3. Highways. People who work with highways use the word grade instead of slope. Grade and slope mean the same thing but grade is usually written as a percentage.

1. If a highway has a grade of 9% at a particular place, what is the slope of the highway?

2. A slope of $-\dfrac{3}{40}$ stands for a downward grade of _____ percent.

3. The steepest parts of Interstate Highways have a grade of 7%. At this rate, how many feet would you climb in a mile (measured horizontally) of driving?

4-5. Climbs and Descents.

4. Some jet aircraft can climb at a slope of 45%. At this rate, how high can the jet get by the time it is 5 km (measured horizontally) from the place it took off?

5. A submarine can dive at a slope of -30%. At this rate, how much horizontal distance is needed to dive 2000 feet below the surface?

416

Lesson 5

The Slope of a Line

If 3 or more points are not on the same line, they determine different slopes.

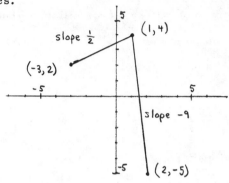

The slope determined by (-3, 2) and (1, 4) is

$\dfrac{4 - 2}{1 - -3}$, which is $\dfrac{2}{4}$ or $\dfrac{1}{2}$.

The slope determined by (1, 4) and (2, -5) is

$\dfrac{-5 - 4}{2 - 1}$ or -9.

But if all points are on the same line, the slopes determined will be the same. Any two points on the segment below will determine a slope of -2.

(0, 3) and (2, -1) determine the slope:

$$\dfrac{-1 - 3}{2 - 0} = \dfrac{-4}{2} = -2$$

(1, 1) and (2, -1) determine the slope:

$$\dfrac{-1 - 1}{2 - 1} = \dfrac{-2}{1} = -2$$

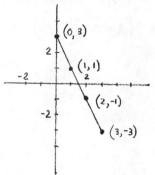

417

This is an important property of lines and slope: On a given line, every pair of points gives the same slope. So we can speak of <u>the</u> slope of a line. The slope of a line can be calculated using any two points on it.

Example: The graph of $x - 2y = 6$ is a line. What is its slope?

Solution: Find two points on $x - 2y = 6$. $(8, 1)$ and $(0, -3)$ will do. Calculate the slope.

$$\frac{-3 - 1}{0 - 8} = \frac{-4}{-8} = \frac{1}{2}$$

So the line has slope $\frac{1}{2}$. This means that it goes up $\frac{1}{2}$ unit for every unit you go to the right.

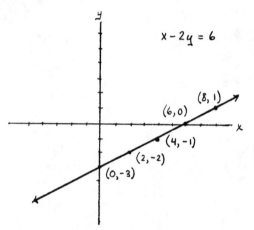

It is easiest to check the slope of a line if there is the <u>same</u> scale on each axis. The dashed line on page 419 has slope 1. It goes up just as fast as it goes across. The dotted line has slope -1. It goes down just as fast as it goes across. Steeper lines have higher slopes. A horizontal line neither goes up nor down, so it has slope 0.

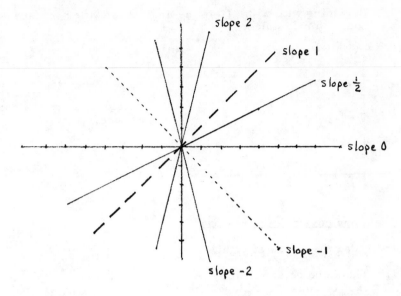

In applications, scales on the axes are almost always different. Still, the slope has a meaning. It is always a rate.

1. Travelling d miles in t hours at 50 miles an hour: $d = 50t$

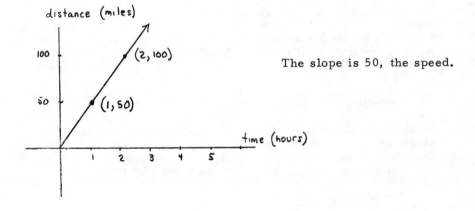

The slope is 50, the speed.

99196

2.   Beginning with $50, L is the amount left after n tapes are bought for $5 apiece.

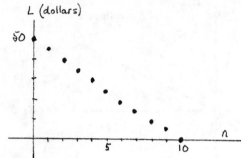

L = 50 - 5n

The slope is -5, the cost per tape.

Questions covering the reading

1-3.   Use the figure at right.

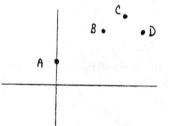

1.   The slope determined by A and B equals the slope determined by B and what other point?

2.   The slope determined by ___ and ___ is negative.

3.   The slope determined by ___ and ___ is zero.

4.   When will points determine the same slope?

5-8.   The graph of the solution set to each equation is a line.   (a) What is the slope of that line?   (b) Check your answer by graphing.

5. $3x + 2y = 6$ 6. $y = 4x - 3$

7. $2y - x = 7$ 8. $x = y - 2$

9-10.   What is the meaning of the slope when each situation is graphed?

9.   the distance travelled in a given amount of time

10.   the amount L left after n objects are bought

420

11-16. The scales on the two axes are the same. Give the slope of the indicated line.

11. a   12. b

13. c   14. d

15. e   16. f

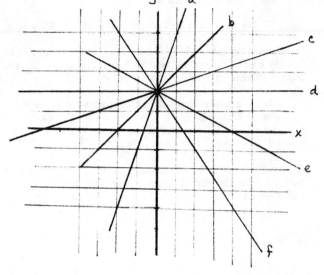

Questions testing understanding of the reading

1-4. Draw a line with the given slope. Use the same scale on each axis.

1. 4   2. $\frac{1}{6}$ 3. 1   4. $-\frac{2}{3}$

5-8. The point (2, 5) is on a line. Name one other point on that line if the slope of the line is:

5. 3   6. -2   7. $\frac{1}{5}$ 8. $-\frac{2}{3}$

9-12. A line goes through the origin. Name one other point on the line if the slope of the line is:

9. 5   10. 2.3   11. -6   12. m

13-16. A situation is given. In each case the graph of the equation is a line. (a) Graph the line. (b) Determine the slope of the line. (c) What does the slope mean?

13. Let C be the cost of n pillows at $8.95 each. Then C = 8.95n.

14. It costs a printer $1000 to print 500 books. He can sell them for $4 each. Let p be his profit and b be the number of books sold. Then p = 4b - 1000. (Let b be the first variable.)

15. A college student is 120 km from home. He decides to bike home and can make 15 km/hr. Let h be the number of hours biked and d be his distance from home. Then d = 120 - 15h. (Let h be the first variable.)

16. A savings account is started with $25. $5 is added each month. Let t be the total amount in the account. Then after m months, t = 25 + 5m. (Let m be the first variable.)

Lesson 6

Slope-Intercept Equations for Some Lines

Here is a question you have seen before.

Question: Suppose you have a jar at home with $12.43 in it. If you put in $.25 a day beginning tomorrow, how much will you have after x days?

Answer: Let y be the amount you will have after x days. Then, as you know,

$$y = 12.43 + .25x$$

So that no one is tempted to incorrectly add the 12.43 and .25, this answer is usually written

$$y = .25x + 12.43$$

When solutions to this equation are graphed, the points all lie on a line. The number .25 is the slope of the line--it is the change in the amount y per day. The number 12.43 signifies where the line crosses the y-axis. It crosses at (0, 12.43). 12.43 is called the y-intercept of the line.

Some points
(0, 12.43)
(1, 12.68)
(2, 12.93)
(3, 13.18)
(4, 13.43)

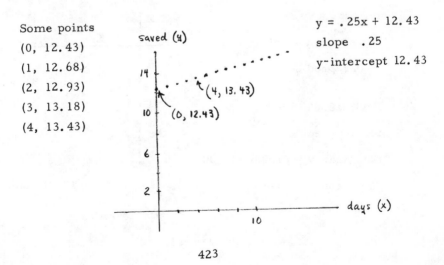

$y = .25x + 12.43$
slope .25
y-intercept 12.43

423

This example is important because it shows:

> An equation for a line is determined
> by knowing the slope and y-intercept.

Many situations are like the saving situation.

Examples:

1. Begin with $20.00. Save $.50
   a day.

   Equation: $y = .50x + 20.00$

   (Graphed dotted)

2. Begin with $20.00. Spend
   $1.50 a day.

   Equation: $y = -1.50x + 20.00$

   (Graphed dashed)

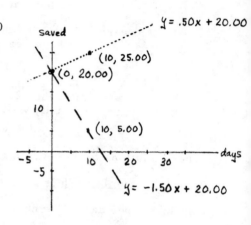

3. Begin $8.00 in debt. Pay it
   off $.50 a week.

   Equation: $y = .50x - 8.00$

   (Graphed dotted)

4. Begin $8.00 in debt. Spend
   $1.00 a week.

   Equation: $y = -1.00x - 8.00$

   (Graphed dashed)

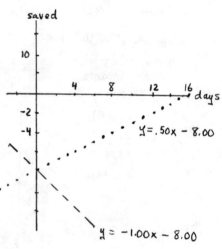

424

Every line which intercepts the y-axis in a single point has an equation like those on page 424. In general,

$$y = \text{(rate saved or lost)} \quad x + \text{(amount started with)}$$

$$y = \qquad \text{slope} \qquad x + \qquad \text{y-intercept}$$

It is customary to let the letter m stand for the slope and b stand for the y-intercept.

> The equation $y = mx + b$ is called the
>
> <u>slope-intercept equation</u> for a line.

Examples: Slope-Intercept equations for lines

1. The line with slope 3 and y-intercept 4 has equation $y = 3x + 4$.

2. An equation for the line which contains $(0, -8)$ with slope $\frac{2}{3}$ is $y = \frac{2}{3}x - 8$.

3. The graph of $y = -10x + 6$ is a line with slope -10 and y-intercept 6.

4. The graph of $y = 3 + \frac{2}{3}x$ is a line with slope $\frac{2}{3}$ and y-intercept 3.

<u>Questions covering the reading</u>

1. Suppose you begin with \$7.62 in a jar and add \$.50 a day. Let y be the amount you will have after x days.

   (a) Give an equation relating x and y.

   (b) If you graph all pairs (x, y), the points lie on a line with slope _____ and y-intercept _____.

2. Repeat Question 1 if you begin with \$30.00 and spend \$1.25 a day.

3-4. Suppose you see the graph of a line.

3. How can you tell the y-intercept?

4. How can you tell the slope?

5-9. Give an equation for the line with:

5. slope 2, y-intercept 3

6. slope $\frac{2}{3}$, y-intercept 6

7. slope -9, y-intercept -1

8. slope $-\frac{1}{5}$, y-intercept 0

9. slope 0, y-intercept $-\frac{2}{3}$

10-14. Graph the lines of Exercises 5-9.

15-16. Consider the line whose points (x, y) satisfy $y = mx + b$.

15. What is the slope of the line?

16. What is the y-intercept of the line?

17-30. Give (a) the slope and (b) the y-intercept of each line.

17. $y = 5x + 4$

18. $y = -3x + \frac{1}{5}$

19. $y = \frac{1}{2}x - \frac{5}{2}$

20. $y = 2(x + 6)$

21. $y = -x + 0$

22. $y = x$

23. $y = 3 - 4x$

24. $y = \frac{3x - 8}{5}$

25. $y = 0 \cdot x + 15$

26. $y = -\frac{1}{2}$

27. $y = \frac{6 + x}{2}$

28. $y = -14 - 13x$

29. $y = 12 - x$

30. $y = \frac{2}{3}x - \frac{1}{3}$

31-34. Give an equation which describes amount saved (y) in terms of number of weeks (x).

31. (a) Begin with $10.00, save $2.00 a week.
    (b) Begin with $10.00, save $1.00 a week.

32. (a) Begin with $20.00, spend $3.00 a week.
    (b) Begin with $20.00, spend $5.00 a week.

33. (a) Begin $15.00 in debt, pay off $2.00 a week.
    (b) Begin $30.00 in debt, pay off $2.00 a week.

34. (a) Begin owing $22.40, spend $.50 a week.
    (b) Begin with no money, spend $.50 a week.

Questions testing understanding of the reading

1-10. Each situation naturally leads to an equation of the form
$y = mx + b$. (a) Tell what x and y could stand for. (b) Give the
equation. (c) Graph all solutions to the equation.

    Example: A plane climbs at 50 feet per second in altitude,
               starting at 22,000 feet.

    Answer: (a) Let x = no. of seconds since the climb began
                  Let y = altitude of plane

            (b) Then $y = 50x + 22,000$

            (c) Three points on the graph are (0, 22000), (1, 22050),
                and (2, 22100).

1. It costs $11 and 10¢ a mile to rent a car.

2. An envelope weighs 8 grams and a sheet of paper weighs 4 grams.

3. A person is 10 kg underweight and plans a diet to gain .2 kg a day.

4. The boat is 3.5 km away and is travelling to us at 4 km an hour.

5. A plane lowers its altitude at the rate of 5 meters per second.
It begins at an altitude of 8500 meters.

6. A bookstore pays $250 for a large collection of old books. It
plans to sell the books for $.75 apiece.

7. A student expects a test score of 83. It is estimated that every
hour of studying will raise the score by 3 points.

8. Tickets for a play are $1.50 each. You have $25 and want to buy
tickets for some friends.

9. Town A has a population of 25,000 and about 100 new people are
coming in each month.

10. Town B has a population of 50,000 and seems to be losing people
at a rate of 50 people a month.

11. Verify that 6 is the slope of $y = 6x - 2$ by finding two points on the line and calculating the slope.

12. Verify that $-\frac{1}{2}$ is the slope of $y = -\frac{1}{2}x + 3$ by finding two points on the line and calculating the slope.

Question for discussion

1. The word "intercept" is used in "y-intercept." It also is used in football--a pass can be intercepted. What do these two uses have in common?

Lesson 7

The Finding of the Fahrenheit-Celsius Conversion Formula

In many earlier lessons you have used a formula connecting Fahrenheit and Celsius temperatures. Here is the way that formula was found.

By international agreement, F and C were assumed to be related by a linear equation. And two key temperatures on each scale were made equal.

Freezing point of water: $F = 32^{\circ}$, $C = 0^{\circ}$

Boiling point of water: $F = 212^{\circ}$, $C = 100^{\circ}$

So the desired line goes through $(32, 0)$ and $(212, 100)$.

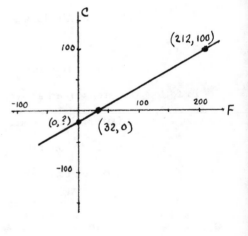

Step 1.   The slope of the line is

$$\frac{100 - 0}{212 - 32}$$

That reduces to $\frac{5}{9}$. So the equation of this line in slope-intercept form is

$$C = \frac{5}{9}F + b$$

where b is to be found. (It's the ? on the graph.)

Step 2.   When F is 32, C is 0. Substituting for F and C:

$$0 = \frac{5}{9} \cdot 32 + b$$

Solving this equation, $b = -\frac{5}{9} \cdot 32$. (You'll see why this isn't multiplied.)

Step 3. Substituting the value just found for b in the equation in step 1.

$$C = \frac{5}{9}F - \frac{5}{9} \cdot 32$$

Applying distributivity, $C = \frac{5}{9}(F - 32)$, the equation we have been using. This equation converts Fahrenheit temperatures into Celsius. It is a conversion formula.

You know that two points determine a line. You usually find that line by using a ruler. That is a geometric way. The above process finds the line in an algebraic way. The algebraic way was developed by Descartes in the early 1600's.

To find an equation for a line through two points, you need only to follow the steps used above.

1. Determine the slope of the line. This gives
    m in the equation $y = mx + b$.

2. Use either of the given points to find b (as
    in Step (2) above).

3. Having found m and b, you have the slope-
    intercept equation for the line.

Example: Find an equation for the line containing (2, 9) and (-4, 6).

Solution: Step 1. Find the slope of the line.

Let m be the slope.

$$m = \frac{6 - 9}{-4 - 2} = \frac{-3}{-6} = \frac{1}{2}$$

So $y = \frac{1}{2}x + b$

Step 2. The line contains (2, 9). Substituting 2 and 9 for x and y:

$9 = \frac{1}{2} \cdot 2 + b$

$9 = 1 + b$

$8 = b$

Step 3. Since $m = \frac{1}{2}$ and $b = 8$, $\underline{y = \frac{1}{2}x + 8.}$ Underlined is the equation for the line.

Check: The two given points must work in the equation. You should check that they do.

The finding of the line through two points is an important skill. An application of this skill to prediction is given in the next lesson.

Questions covering the reading

1-3. Let F and C stand for temperatures in degrees Fahrenheit and Celsius.

1. What 2 pairs of temperatures are the basis for finding the formula connecting F and C?

2. What equation relates F and C?

3. When $F = 0°$, $C = $ _____.

4. What is a geometric way of finding the line through two points?

5. In the algebraic way of finding the line through two points, describe the:

   (a) first step   (b) second step   (c) third step

431

6-8. A line has equation $y = 5x + b$. Find b if the line also contains the given point.

6. $(2, 3)$ 7. $(-6, 1)$ 8. $(\frac{2}{5}, 7)$

9-11. A line has equation $y = -\frac{2}{3}x + b$. Find b if the line also contains the given point.

9. $(0, 4)$ 10. $(-2, 5)$ 11. $(6, 8)$

12-21. Find an equation for the line through the two given points.

12. $(1, 2)$, $(7, 9)$ 13. $(30, 11)$, $(25, 26)$

14. $(6, -3)$, $(8, -9)$ 15. $(-1, 10)$, $(-4, 4)$

16. $(-6, 1)$, $(2, -4)$ 17. $(6, -5)$, $(0, 3)$

18. $(\frac{1}{2}, 0)$, $(\frac{3}{2}, \frac{5}{3})$ 19. $(1.4, 3.8)$, $(2.9, -4.7)$

20. $(6, 9.40)$, $(14, 11.40)$ 21. $(1960, 5)$, $(1975, 20)$

22. Suppose you forgot the boiling point of water but remembered that $20^{\circ}C = 68^{\circ}F$ (room temperature). Using this and $0^{\circ}C = 32^{\circ}F$, find a formula relating Fahrenheit and Celsius temperatures.

Questions applying the reading

1. Chirping crickets. Biologists have found that the number of chirps some crickets make per minute is related to the temperature. The relationship is very close to being linear. (You could determine this relationship by going out on days with two different temperatures.) At $68^{\circ}F$ the crickets chirp about 124 times a minute. At $80^{\circ}F$ they chirp about 172 times a minute. Find the linear equation relating Fahrenheit temperature F and number of chirps c. (Hint: Begin with two ordered pairs suggested by the given information.)

C 2. Snakes. It was reported in 1960 that the total length y and tail length x of females of the snake species lampropeltis polyzona are closely related by a line. When x = 60mm, y = 455mm. When x = 140mm, y = 1050mm. (These snakes have total length from 200mm to 1400mm.) Find the linear relationship between x and y.

3. Stretching springs. If a weight stretches a spring, the amount

of weight w (for small weights) is linearly related to the length L of the spring. Suppose a weight of 400 grams stretches a spring 30mm. What is a relationship between w and L? (Hint: A weight of 0 grams doesn't stretch a spring.)

C 4. Expanding rods. A copper rod stretches with the heat. Suppose a rod is 38.22cm long at 15°C and 38.28cm long at 100°C. Find a linear relationship between length and temperature. (This relationship is good for temperatures under 150°C.)

5. The games of the 21st modern Olympiad were in 1976. The games of the 20th Olympiad were four years earlier. Let Y be the year of the nth summer Olympic games. (a) Give the linear formula which relates n and Y. (b) Use this formula to determine when the 1st modern Olympic games were held.

6. Every two years a new "Congress" is elected in the United States. The 1st Congress was in 1789. So the 2nd Congress was in 1791. (a) Find a formula which relates the number of the Congress and the year. (b) Use the formula to figure out which Congress is presently in session.

Question for discussion

1. How can you tell that the formulas in Questions 5 and 6 above must be linear?

Lesson 8

The Evolution of the Mile Record

In 1875, Walter Slade of Great Britain ran a mile in 4 minutes, 24.5 seconds. At that time, this was the fastest anyone had ever been timed. Now even high school students often do better. Here is how the record has evolved in the years since 1875.

1875	Walter Slade, Britain	4:24.5
1880	Walter George, Britain	4:23.2
1882	George	4:21.4
1882	George	4:19.4
1884	George	4:18.4
1894	Fred Bacon, Scotland	4:18.2
1895	Bacon	4:17.0
1911	Thomas Connett, U.S.	4:15.6
1911	John Paul Jones, U.S.	4:15.4
1913	Jones	4:14.6
1915	Norman Taber, U.S.	4:12.6
1923	Paavo Nurmi, Finland	4:10.4
1931	Jules Ladoumegue, France	4:09.2
1933	Jack Lovelock, New Zealand	4:07.6
1934	Glenn Cunningham, U.S.	4:06.8
1937	Sydney Wooderson, Britain	4:06.4
1942	Gunder Haegg, Sweden	4:06.2
1942	Arne Andersson, Sweden	4:06.2
1942	Haegg	4:04.6
1943	Andersson	4:02.6
1944	Andersson	4:01.6
1945	Haegg	4:01.4
1954	Roger Bannister, Britain	3:59.4
1954	John Landry, Australia	3:58.0
1957	Derek Ibbotson, Britain	3:57.2
1958	Herb Elliott, Australia	3:54.5
1962	Peter Snell, New Zealand	3:54.4
1964	Snell	3:54.1
1965	Michel Jazy, France	3:53.6
1966	Jim Ryun, U.S.	3:51.3
1967	Ryun	3:51.1
1975	Filbert Bayi, Tanzania	3:51.0
1975	John Walker, New Zealand	3:49.4

Evolution of the Mile Record 1875-1975

435

The points on the graph lie very close to a line. This surprises people who think that the times would "level off." But they haven't yet. In fact, the record has gone down faster since 1935 than it did before then. Look at the graph to see that this is so.

Because the data is nearly on a line, you can find an equation which estimates what the record is in a given year. This equation can be used to predict what might happen in the future. Here is how this is done.

Since 1900, a line which fits the data well goes through (1900, 4:20) and (1975, 3:50). (Use the edge of a sheet of paper to check this.) Converting the times to seconds, the points are (1900, 260) and (1975, 230). The slope of the line is $-\frac{30}{75}$ or $-.4$. Using the method of Lesson 7, an equation for the line is

$$\text{Time in seconds} = -.4 \cdot \text{Year} + 1020$$

$$t = -.4Y + 1020$$

To get the time in 1985, substitute 1985 for Y.

$$t = -.4 \cdot 1985 + 1020$$

$$= 226$$

So a record of 226 seconds, or 3 minutes, 46 seconds, might be expected.

In the year 2000, using this equation

$$t = -.4 \cdot 2000 + 1020$$

$$= 220$$

The record might be expected to be 3 minutes, 40 seconds.

In the year 2550, at this rate

$$t = -.4 \cdot 2550 + 1020$$

$$= 0$$

The record would be 0 seconds! This shows that the line could not

fit the data forever.

Questions covering the reading

1-5. Give the world record for the mile at the end of

1.   1875   2.   1900   3.   1925   4.   1950   5.   1975

C   6-10.   According to the line which fits the data well (through (1900, 4:20) and (1975, 3:50)) what will be the world record for the mile in:

6.   1985   7.   1990   8.   2075   9.   1950   10.   2975

11.   How does your answer to Question 9 compare with the actual record?

12.   What does the answer to Question 10 show?

C   13-17.   Another line which fits the data well goes through (1875, 4:24) and (1975, 3:51).

13.   Find an equation for this line.

14.   According to this equation, what will be the record in 1985?

15.   According to this equation, what will be the record in 2000?

16.   According to this equation, when would the record be 0 seconds?

17.   According to this equation, when would the record be 3 minutes?

Questions extending the reading

1.   One of the most famous uses of extrapolation was by the French scientist Jacques Charles in 1787. He noticed that gases expand when heated, contract when cooled. (This can be verified by

blowing up a balloon and then putting it in the freezer compartment of a refrigerator. The balloon will shrink.)

(a) Suppose a particular gas had a volume of 500 cc at 27°C and a volume of 605 cc at 90°C. Charles used this kind of data to get a linear equation which related volume and temperature. (He had more points and noticed they seemed to lie on a line.) Repeat what Charles did.

(b) Use the equation you got in Part (a). Find out what temperature gives a volume of 0 cc. By doing this you have estimated the coldest temperature possible. (This temperature is called <u>absolute zero</u>. It was first approximated by Charles.)

2. A line contains (2, 3) and (11, 6).

    (a) Find an equation for the line.

    (b) Use the equation to determine what y is when x is 10.

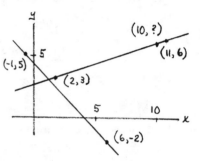

3. A line contains (-1, 5) and (6, -2).

    (a) Find an equation for the line.

    (b) Use the equation to determine where the line crosses the x-axis.

4. Suppose it costs $3.50 for a 14-inch pizza and $2.00 for a 10-inch pizza.

    (a) Find a linear equation relating pizza diameter and cost.
    (b) What might a 12-inch pizza cost?
    (c) What might a 17-inch pizza cost?
    (d) Why would you be getting a good deal if you bought a pizza for the price found in part (c)?

5. According to the September, 1970, issue of <u>Changing Times</u>, it cost then about $96 per month to drive a car 200 miles per month. It cost about $128 per month to drive a car 1000 miles per month. (These costs would be higher now.) Use the ordered pairs (200, 96) and (1000, 128).

    (a) Get a linear equation relating miles driven to cost per month.
    (b) What does the slope of this line stand for?
    (c) What does the y-intercept of this line mean?
    (d) Predict the cost if you drive 500 miles per month.
    (e) Predict the cost if you drive 1500 miles per month.

6. The U.S. population was about 180 million in 1960 and 205 million in 1970.

   (a) Find a linear equation relating date and population.
   (b) Use this equation to estimate the population in 1963.
   (c) Use this equation to estimate what the population was in 1925.
   (d) Use this equation to estimate the population in 2025.

7. Estimates like those of Question 6 are usually not particularly accurate. Why not?

Lesson 9

Horizontal and Vertical Lines

Every point on the line at
right has y-coordinate 3. So an
equation for the line is

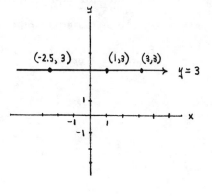

$$y = 3.$$

It is easy to convert this equation
to slope-intercept form. Since the
slope is 0 and y-intercept is 3,

$$y = 0 \cdot x + 3.$$

So the slope-intercept form can be used with <u>horizontal</u> lines. Every
horizontal line has an equation of the form $y = 0 \cdot x + b$. Simplifying
the right side, $y = b$.

A <u>vertical</u> line (like the one at
right) does not "go across." So it
is impossible to calculate a slope
for this line.   Look what happens
when you try.

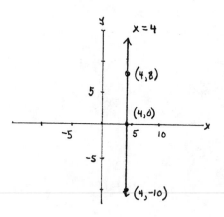

Pick two points on the line,
let's say (4, 8) and (4, -1).
Try to calculate slope.

$$\frac{-1 - 8}{4 - 4}$$

You get 0 in the denominator.
This expression does not
stand for a real number.

440

So vertical lines do not have a slope. And there cannot be a slope-intercept equation for a vertical line. But for every point on the line, <u>the x-coordinate is 4</u>

This is abbreviated: <u>x = 4</u>

Every other vertical line has a similar equation.

To summarize:

The horizontal line containing $(0, b)$ has equation $y = b$. It has slope 0 and slope-intercept equation $y = 0x + b$.	The vertical line containing $(a, 0)$ has equation $x = a$. It has no slope and no slope-intercept equation.

A line which is not horizontal or vertical is called <u>oblique</u>. All oblique lines have slope-intercept equations.

<u>Questions covering the reading</u>

1-4. Give the equation of the vertical lines through:

1. $(-2, 5)$ 2. the origin   3. $(a, 0)$ 4. $(-40, 72)$

5. Why do vertical lines lack a slope?

6-9. Give an equation for the horizontal line through:

6. $(9, 12)$ 7. $(0, b)$ 8. the origin   9. $(7.5, -3.2)$

10. What can be said about slope and horizontal lines?

11-16. (a) Tell whether the line is horizontal, vertical, or oblique.
(b) Name two points on the line. (c) If the line has a slope, calculate the slope.

11. $x = -3$ 12. $x = 0$ 13. $y = 2$

14. $x + y = 2$ 15. $y = x$ 16. $y = -9$

17. $y = 3x$ 18. $x = 4y$ 19. $x = \frac{1}{2}$

Questions testing understanding of the reading

1-6. Give an equation for the:

1. vertical line containing $(2, -9)$.

2. vertical line through $(.03, 3.6)$.

3. horizontal line through $(-2.3, -1)$.

4. horizontal line containing $(7000, 4926)$.

5. line through $(2, 5)$ and $(2, 40)$.

6. line through $(7, 0)$ and $(2.5, 0)$.

7-8. Give the point of intersection of the lines with equations:

7. $x = 4$ and $y = 9$ 8. $x + 6 = 0$ and $y = 0 \cdot x + 9$

9. A person weighs 70 kg on February 1st and 70 kg on February 28th. Calculate the average change of weight per day. What does this have to do with slope? What does this have to do with horizontal or vertical lines?

10. On March 1st a person is 150 cm tall. At the same time, the same person is 225 cm tall according to another measurement. What can you say about the rate of change in height per day? What does this have to do with horizontal or vertical lines?

442

Lesson 10

Equations for All Lines

You have now studied lines for some time. So you may wonder: <u>Why are lines studied</u>? There are many reasons. First, the line is the most basic geometric figure in mathematics. Triangles, polygons, networks, and designs can be broken up into lines and segments. Second, lines can be used to approximate curves.

Third, any quantity which increases by a constant amount leads to line graphs. Telephone rates, profit, distances, and many other examples have been given. Fourth, other real data (like the mile record) seems to lie along lines. Fifth, many measurements use lines. Think of rulers, thermometers, and most other measuring devices.

<u>What else can you study about lines</u>? There is much. In this course you will study situations in which more than one line is involved. You will also study regions determined by lines. In geometry, you will study geometric properties of lines. These include parallelism, perpendicularity, and lots of figures involving lines. Linear independence and linear transformations are some ideas you might have in later courses.

Equations for lines are simple. This makes it helpful to be able to find linear relationships between variables. People who study or use statistics measure the closeness of data to a line by a number called the linear correlation coefficient between the two variables. Linear regression is an idea used when two or more variables are present.

All of this means that if you study more mathematics or more applications of mathematics, you are bound to meet lines again and again.

All this makes it useful to have a simple single way of describing all lines. Remember that every non-vertical line has an equation of the form $y = mx + b$. Add ^-mx to each side. Then you get

$$^-m \cdot x + 1 \cdot y = b$$

Every vertical line has an equation of the form $x = a$. This is a shorter way of writing

$$1 \cdot x + 0 \cdot y = a$$

Putting all this together: This shows that every line has an equation of the form

$$A \cdot x + B \cdot y = C$$

where you usually know A, B, and C.

Examples: Ax + By = C form

1. The equation $x = 10$ (of a vertical line) is equivalent to $1 \cdot x + 0 \cdot y = 10$. Here $A = 1$, $B = 0$, $C = 10$.

2. The slope-intercept equation $y = \frac{1}{2}x - 35$ is equivalent to $-\frac{1}{2} \cdot x + 1 \cdot y = -35$. Here $A = -\frac{1}{2}$, $B = 1$, $C = -35$.

If you have an equation like

$$2x + 3y = 18$$

you can convert it to slope-intercept form by solving for y.

$$3y = -2x + 18$$
$$y = \frac{-2x + 18}{3}$$
$$y = -\frac{2}{3}x + 6$$

So its slope is $-\frac{2}{3}$ and its y-intercept is 6. This tells you that, when $B \neq 0$, the equation $Ax + By = C$ leads to a line.

To summarize: When you graph all of the solutions to an equation in this form, you will get a line. So we now have a way of telling when solutions to a sentence will lie on a line.

Every line in the Cartesian coordinate plane has an equation of the form $Ax + By = C$ and every equation of this form has a graph which is a line.

Each of the equations at right is of the form $Ax + By = C$. So the graph of the solutions of each is a line.

$$3x + 2y = 5.5$$
$$y = 9$$
$$x - 9y = 11$$
$$x = -2.3$$

Questions covering the reading

1-8. Each equation is of the form $Ax + By = C$. (a) Tell what A, B, and C are. (b) If possible, solve the equation for y. (c) If the line has a slope, tell what the slope is.

1. $4x + 2y = 5$

2. $3x - 6y = 30$

3. $x - 8y = 2$

4. $\frac{1}{2}x + 0y = 10$

5. $9x - \frac{1}{3}y = \frac{2}{3}$

6. $-4y = 20$

7. $2x + y = 8$

8. $x + y = 7$

9-10. **Multiple choice.** Which is not an equation for a line?

9. (a) $x+y = 3$ (b) $y - x = 3$ (c) $x = 3$ (d) $xy = 3$

10. (a) $\frac{2x}{y} = 10$ (b) $y = 5x$ (c) $5x = 3y$ (d) $0x = 3$

11. Give four reasons why lines are studied.

12. Name two things you will study about lines if you take more mathematics courses.

13-22. Put each equation into $Ax + By = C$ form and tell what A, B, and C are.

13. $y = 4x + 5$

14. $y = 3x - 2$

15. $2y = x - 8$

16. $\frac{2}{3}y = 5(x + 6)$

17. $x = 2(y - 4)$

18. $x = 9$

19. $y = 6.32$

20. $1 = x - y - 5$

21. $\frac{x + 2}{3} = y$

22. $5y + 6 = 2x - 8$

Questions for discussion and exploration

1-4. Use a dictionary, encyclopedia, or other reference book. Find out something about the given idea.

1. linear programming

2. linear independence

3. linear regression

4. linear correlation

446

5. Here are the results of a historical study of the actual number
    of German submarines sunk by U.S. ships during 16 months of
    World War II. (Data is taken from Probability with Statistical
    Applications, by Mosteller, Rourke, and Thomas, Addison-Wesley
    Publishing Co., 1970.)

Month	Actual number x sunk (by German reports)	U.S. reported sinking y (in Navy reports)
1	3	3
2	2	2
3	6	4
4	3	2
5	4	5
6	3	5
7	11	9
8	9	12
9	10	8
10	16	13
11	13	14
12	5	3
13	6	4
14	19	13
15	15	10
16	15	16

On p. 448 is a graph of the 16 ordered pairs (x, y). Numbers by
the points are month numbers.

5. (continued)

(a) Guess at the equation of a line which most closely fits the data. (Finding this line is part of linear regression theory.)

(b) Did the U.S. Navy overestimate or underestimate the number of submarines it had sunk?

(c) If the U.S. Navy had been perfectly accurate in its reports, what would have been an equation for the line containing all points?

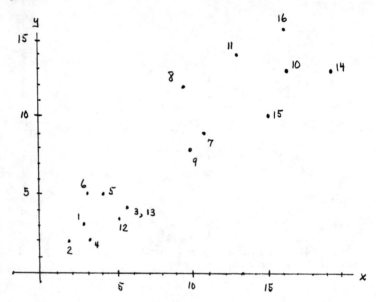

Chapter Summary

Real situations lead to many different types of graphs. Some are discrete. Some are continuous. Some of these graphs can be described by simple sentences. For example, every line has an equation of the form $Ax + By = C$. That is, given the numbers A, B, and C, the set of ordered pair (x, y) solutions to $Ax + By = C$ is a line.

When some data is graphed, the points lie very near a line. This is the case with the record time t (in seconds) for the mile run. Over the past 75 years, if Y is the year, then $t = -.4Y + 1020$.

The $-.4$ in the above equation is the slope or rate of change of the line. It indicates that, on the average, the mile record has been lowered .4 seconds a year. In general, the slope determined by (x_1, y_1) and (x_2, y_2) is $\dfrac{y_2 - y_1}{x_2 - x_1}$. On a given line, this slope will always be the same. The line with slope m and y-intercept b has the slope-intercept equation

$$y = mx + b$$

449

For a horizontal line, the slope m is 0. So the horizontal line through (0, b) has equation

$$y = b.$$

For a vertical line, slope is not defined. The vertical line through (a, 0) has equation x = a.

There is only one line through two points. Knowing the two points, an equation can always be found. For example, given the two points $(32^{\circ}, 0^{\circ})$ and $(212^{\circ}, 100^{\circ})$ of the Fahrenheit-Celsius (F, C) line,

$$C = \frac{5}{9}(F - 32)$$

Of course, there are situations in which data does not lie on or close to a line. Slopes can still be calculated but will change, depending upon the points chosen.

CHAPTER 10

POWERING

Lesson 1

The Repeated Multiplication Model of Powering

Some banks pay savers 5% per year interest on the money saved. This means that P dollars saved will bring .05P dollars interest.

$$\text{The total} = P + .05P \text{ dollars}$$

$$= 1.05P \text{ dollars}$$

The number 1.05 is a scale factor which tells what the money is multiplied by each year. If the interest is kept in the account, there will be

after this much time	in the account
1 year	$1.05P$
2 years	$1.05 \cdot 1.05P$
3 years	$1.05 \cdot 1.05 \cdot 1.05P$
$\vdots$	$\vdots$
n years	$\underbrace{1.05 \cdot 1.05 \cdot \ldots \cdot 1.05P}_{\text{n factors}}$

The repeated multiplication with 1.05 used n times is abbreviated

$$1.05^{n}$$

and called   "the nth power of 1.05"

or   "1.05 to the nth."

So after n years there would be $(1.05)^{n} \cdot P$ in the account.

451

This repeated multiplication leads to an operation called <u>powering</u>.

<u>Repeated Multiplication
Model
of Powering</u>:

> Let n be a positive integer and
> B be any real number. Then
> $\underbrace{B \cdot B \cdot \ldots \cdot B}_{n \text{ factors}} = B^n$, the <u>nth power of B</u>.

In the power B^n we call B the <u>base</u> and n the <u>exponent</u>. The exponent is always written to the right and above the base.

Examples: Calculating some powers with 1.05 as base.

1. $(1.05)^1 = 1.05$ The parentheses in $(1.05)^1$ and $(1.05)^2$ are not necessary

2. $(1.05)^2 = 1.05 \cdot 1.05 = 1.1025$

3. $1.05^3 = 1.05 \cdot 1.05 \cdot 1.05$

    $= 1.1025 \cdot 1.05 = 1.157625$

4. $1.05^4 = 1.05 \cdot 1.05 \cdot 1.05 \cdot 1.05$

    $= 1.157625 \cdot 1.05 = 1.21550625$

5. If you deposited \$600 at 5% yearly interest and kept the interest in the account, you would have at the end of 4 years:

    $$(1.05)^4 \cdot \$600$$

    $$= 1.21550625 \cdot \$600$$

    $$\approx \quad \$729.30$$

There are many applications of powering. So the calculation of powers is a very important skill. For powering more than any other operation, an electronic calculator can be helpful. But some powers are easily calculated by hand.

6. $2^5 = 2 \cdot 2 \cdot 2 \cdot 2 \cdot 2 = 32$

7. $(-2)^5 = -2 \cdot -2 \cdot -2 \cdot -2 \cdot -2 = -32$

To "calculate" a power means to write it as a decimal. Calculated, $(1.06)^{10} = 1.7908\dots$. If you do not have a calculator, you should not attempt to calculate $(1.06)^{10}$. It is best to keep that number written as a power.

Questions covering the reading

1. Multiple Choice. $x^2 =$

(a) $2x$ (b) $x + x$ (c) $x \cdot x$ (d) x_2 (e) none of these

2. Multiple Choice. $6^4 =$

(a) 24 (b) 4^6 (c) 6_4 (d) $6 \cdot 6 \cdot 6 \cdot 6$ (e) none of these

3-6. Consider the expression B^n.

3. What is n called?   4. What is B called?

5. What is B^n called?   6. How is B^n calculated?

7. Suppose you invest \$50 at 6% yearly interest, and keep the interest in the account.
   (a) After 1 year you will have _____.
   (b) After 2 years you will have _____.
   (c) After 10 years you will have _____.
   (d) After n years you will have _____.

8. Suppose you save \$100 at 5% yearly interest. If the interest is kept in the account, how much will you have:

(a) after 1 year.   (b) after 2 years.

(c) after 3 years.   (d) after t years.

453

9-24. Write without using exponents.

9. $(\frac{1}{2})^3$ 10. 7^4 11. $(21.3)^2$ 12. $(\frac{2}{5})^2$

13. $(1.07)^2$ 14. $(1.07)^3$ 15. $(1.07)^4$ 16. 1^{100}

17. $(.2)^2$ 18. $(.3)^2$ 19. $(-5)^2$ 20. $(-5)^3$

21. 6^1 22. 2^6 23. $(\frac{1}{2})^6$ 24. $(1.42)^1$

25. What is the repeated multiplication model of powering?

Questions testing understanding of the reading

1. Calculate 2^1, 2^2, 2^3, 2^4, 2^5, 2^6, 2^7, 2^8, 2^9, and 2^{10}.

2. Calculate 3^1, 3^2, 3^3, 3^4, 3^5, and 3^6.

3. The number x^2 is often called "x squared." How do you think this name arose?

4. The number y^3 is often called "y cubed." How do you think this name arose?

5-10. Place a $<$, $=$, or $>$ sign in the blank. Be careful. Few students get all of these correct.

5. 3^5 ____ 5^3 6. $(\frac{2}{3})^{11}$ ____ $(\frac{2}{3})^{12}$

7. 4^7 ____ 4^6 8. 11^4 ____ 10^4

9. $(.91)^5$ ____ $(.93)^5$ 10. $(-2)^6$ ____ $(-2)^7$

C 11. Calculate what you would have if you invested (keeping interest in the account) \$150 at 5% yearly interest:

   (a) for 1 year.   (b) for 2 years.

   (c) for 3 years.   (d) for n years.

C 12. Repeat Question 11 if \$50 is saved at 8% yearly interest.

C 13. Repeat Question 11 if \$50 is saved at 3% yearly interest.

C 14. Repeat Question 11 if \$1000 is invested at 10% yearly interest.

15-18. Calculate x^1, x^2, x^3, x^4, x^5, and x^6 when x is:

15. -2   16. 1   17. -1   C 18. .9

19. <u>Multiple Choice</u>. $a^b = b^a$

    (a) always   (b) sometimes   (c) never

Lesson 2

The Permutation Model of Powering

Since some powering is short for repeated multiplication, applications of multiplication lead to applications of powering.

Examples: Counting Problems

1. How many different 4-letter words are possible in English?

___ ___ ___ ___

Answer: Each of the 4 slots can be filled in 26 ways. Using the ordered pair model of multiplication, the number of possible words is

$$26 \cdot 26 \cdot 26 \cdot 26 = 456,976$$

Of course, $26 \cdot 26 \cdot 26 \cdot 26 = 26^4$.

2. A coin is tossed 10 times. One possible result is HHTHTTTHHHT. How many different sequences of H's and T's are possible?

___ ___ ___ ___ ___ ___ ___ ___ ___ ___

Answer: Each slot can be filled 2 ways. So there are

$\underbrace{2 \cdot 2 \cdot 2 \ldots \cdot 2}_{10 \text{ factors}}$, or 2^{10} possible sequences. As a decimal,

$2^{10} = 1024$. This is a larger answer than most people would guess.

The above examples are instances of the permutation model of powering. This model generalizes the ordered pair model of

456

multiplication you studied earlier.

Permutation
Model
of Powering:

If each of n blanks can be filled
in r ways, then r^n different
sequences are possible.

Suppose there are 2 blanks ___ ___ and each blank can be
filled in 6 ways. We call the fillers 1, 2, 3, 4, 5, 6. Then here
are the possible sequences.

11	12	13	14	15	16
21	22	23	24	25	26
31	32	33	34	35	36
41	42	43	44	45	46
51	52	53	54	55	56
61	62	63	64	65	66

There are 36 or 6^2 sequences possible, as predicted by the permutation model.

This type of counting is often used to calculate probabilities. If
you toss 1 red and 1 white fair die, the probability of a 5 on the red
and a 6 on the white is $\frac{1}{36}$ or $\frac{1}{6^2}$, because 6^2 outcomes are possible.

Similarly, imagine tossing a fair coin 10 times. There are 2^{10}
possible sequences (from Example 2 on page 456). So each sequence
has probability $\frac{1}{2^{10}}$. So getting HHHHHHHHHH (10 heads) has
probability $\frac{1}{2^{10}}$.

457

Questions covering the reading

1. Suppose blank A can be filled in 10 ways and blank B can be filled in 10 ways.

$$\overline{\quad}\ \overline{\quad}$$
$$\text{A}\quad\text{B}$$

   How many ways are there of filling the two blanks together?

2. Repeat Question 1 if each blank can be filled in 15 ways.

3. Explain why there are 26^4 possible 4-letter words in English. (Of course, some of these would be hard to pronounce.)

4. How many 3-letter words are possible in English?

5. The Hawaiian language has 12 letters: A, E, I, O, U, H, K, L, M, N, P, and W. How many 2-letter words are possible in Hawaiian? (See Question 2, p. 461.)

6. Refer to Question 5. How many 5-letter words are possible in Hawaiian?

7. A coin is tossed 10 times. Write down one of the sequences of H's and T's which are possible. How many different sequences are possible?

8. A coin is tossed 3 times. Write down all of the possible sequences of H's and T's which are possible.

9. How many different sequences of H's and T's are possible if a coin is tossed:

   (a) 1 time　　　　　　(b) twice

   (c) 3 times　　　　　(d) 4 times

   (e) 5 times　　　　　(f) 8 times

   (g) 50 times　　　　(h) n times

10-19. You toss a fair coin. What is the probability of getting:

10. 2 heads in 2 tosses.　　　11. 2 tails in 2 tosses.

12. 3 heads in 3 tosses.　　　13. 3 tails in 3 tosses.

14. 4 heads in 4 tosses.　　　15. 4 tails in 4 tosses.

16. 7 heads in a row.   17. 7 tails in a row.

18. n heads in a row.   19. t tails in a row.

20. How many different sequences of T's and F's are possible in a True-False test with:

    (a) 1 question   (b) 2 questions

    (c) 3 questions   (d) 4 questions

    (e) 5 questions   (f) 15 questions

    (g) 40 questions   (h) q questions

21. Repeat Question 20 if the test contains only always-sometimes-never questions, and possible answers are A, S, or N.

22. Suppose you take a 3-question True-False test and guess at each answer. What is the probability that you will get all 3 questions correct?

23. Suppose you take a 5-question Always-Sometimes-Never test and guess at each answer. What is the probability that you will get all 5 questions correct?

24. You guess at 2 multiple choice questions, each with 5 choices. What is the probability that you will get both correct?

25. What is the permutation model for powering?

Questions extending the reading

1-11. These questions make use of a different extension of the ordered pair model for multiplication, but are related to the permutation model given in this lesson. This extension is the following: If blank 1 can be filled in w_1 ways, and blank 2 in w_2 ways, and blank 3 in w_3 ways, etc., then there are $w_1 \cdot w_2 \cdot w_3 \ldots$ possible sequences with all the blanks filled.

1. A restaurant has 6 different appetizers, 10 different main courses, and 9 different desserts. How many different meals could you eat at this restaurant before you would have to repeat?

2. President Franklin Roosevelt was known as FDR. President Kennedy was known as JFK. President Lyndon Johnson was often called by his initials LBJ. How many different sequences of three initials are possible in the English language?

3. Your telephone area code consists of three digits.

$$\overrightarrow{} \qquad \overset{\uparrow}{} \qquad \overset{\nwarrow}{}$$

any digit from   0 or 1   any digit from
2 to 9   2 to 9

How many different area codes are possible?

4. In some states, license plates contain two letters followed by 1 to 4 numbers. How many different license plates can there be in these states?

5. In some states, license plates contain three letters followed by 1 to 3 numbers. Make up a reasonable question and answer it.

6-7. There are 3 routes joining city W to city X, 5 routes joining X to Y, 4 routes joining Y to Z, and 3 routes joining Z to W. A diagram of the situation is given.

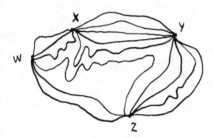

6. In how many ways can a person travel from W to Z passing through X and Y?

7. How many routes are there from Z to W to X to Y?

8. Fill blank A with one of the numerals 1, 2, or 3. Fill blank B with 4 or 5. Fill blank C with 6 or 7. Fill blank D with 8 or 9. (a) Write down all 4-digit numbers which you can get in this way.

$$\overline{\text{A}} \quad \overline{\text{B}} \quad \overline{\text{C}} \quad \overline{\text{D}}$$

(b) How many numbers can be written?

9. The Girls Gym Committee consists of one sophomore, one junior, and one senior. The following girls run for office.

460

Sophomores	Juniors	Seniors
Alicia Alvarez	Phyllis Fairchild	Mary Martell
Ruth Rosen	Holly Hinman	Wanda Wong
	Judy Janowicz	
	Teresa Thomas	

List all of the possible committees which could arise.

10. A school has 500 freshmen, 450 sophomores, 400 juniors, and 350 seniors. In how many ways could a TV station choose 4 students, 1 from each class, for interviews?

11. If blank E can be filled in w_1 ways, blank F can be filled in w_2 ways, and blank G can be filled in w_3 ways, in how many ways can the 3 blanks be filled together?

$$\overline{\text{E}} \quad \overline{\text{F}} \quad \overline{\text{G}}$$

12-17. Review. If $W_1 = 3$, $W_2 = 4$, and $W_3 = 2$, and $W_4 = 6$, calculate:

12. $W_1 W_2 W_3 W_4$

13. $W_3 W_2 W_4 W_1$

14. $W_2 W_3 W_1$

15. $8W_1 + W_2$

16. $10W_1 W_4$

17. $(W_4 W_3) W_1$

Question for discussion

1. Look in a dictionary for the meaning of the word "permutation." (This will help explain why the model of this lesson is called the permutation model.)

Lesson 3

Powering and Order of Operations

What is the meaning of $3 \cdot 2^5$? Without rules for order of operations, you could not know whether to do the multiplication first (that would be 6^5, which is 7776) or the powering first (that would be $3 \cdot 32$, which is 96).

The rule for working with powers is very simple. Powers are calculated before all other operations. This means that operations are done in the following order:

      1. Work inside parentheses.

      2. Calculate powers, from left to right.

      3. Multiply or divide, from left to right.

      4. Add or subtract, from left to right.

Examples: Order of operations using powers

1. $\begin{aligned} 3 \cdot 2^5 &= 3 \cdot 2 \cdot 2 \cdot 2 \cdot 2 \cdot 2 \\ &= 3 \cdot 32 \\ &= 96 \end{aligned}$

2. $\begin{aligned} 4(7 - 2)^2 + 1^3 &= 4 \cdot 5^2 + 1^3 \\ &= 4 \cdot 25 + 1 \\ &= 100 + 1 \\ &= 101 \end{aligned}$

3. If $x = 6$, then $\begin{aligned} 2x^3 - 3x^2 &= 2 \cdot 6^3 - 3 \cdot 6^2 \\ &= 2 \cdot 216 - 3 \cdot 36 \\ &= 432 - 108 \\ &= 324 \end{aligned}$

From Lesson 1, you worked on what is called <u>compound interest</u>. If P dollars are invested at r% yearly interest for n years (and interest is left in the account), then the total amount is

after 1 year: $P + P \cdot \frac{r}{100}$ or $P(1 + \frac{r}{100})$

after 2 years: $P(1 + \frac{r}{100})^2$

after 3 years: $P(1 + \frac{r}{100})^3$

after n years: $P(1 + \frac{r}{100})^n$

For example, if $300 is invested at 5% for 6 years, there is

$$300(1 + \frac{5}{100})^6$$

at the end. This is calculated in the following order.

Work inside parentheses. The number 1.05 is the scale factor. $= 300(1.05)^6$

Calculate power--a calculator is very helpful for this. $\approx 300(1.3401)$

Multiply. $= 402.03$

There is $402.03 at the end of the 6 years.

Now we consider some opposites. To calculate -2^6, think of -2 as $-1 \cdot 2$. So the power is calculated first.

$$-2^6 = -1 \cdot 2^6$$

$$= -1 \cdot 2 \cdot 2 \cdot 2 \cdot 2 \cdot 2 \cdot 2 \quad = -64$$

But to calculate $(-2)^6$, the power is of the entire number.

$$(-2)^6 = -2 \cdot -2 \cdot -2 \cdot -2 \cdot -2 \cdot -2 = 64$$

That is, $-2^6 \neq (-2)^6$. <u>Parentheses are needed to show a power of a negative number.</u>

Examples: Order of operations and negatives.

4. $-6^3 = -216$

5. $(-6)^3 = -6 \cdot -6 \cdot -6 = -216$

6. $-(\frac{1}{10})^2 = -\frac{1}{100}$

7. $(-\frac{1}{10})^2 = -\frac{1}{10} \cdot -\frac{1}{10} = \frac{1}{100}$

8. $-x^4$ cannot be simplified

9. $(-x)^2 = -x \cdot -x = x \cdot x = x^2$

Questions covering the reading

1-12. Calculate:

1. $3 \cdot 2^3$

2. $3 - 2^3$

3. $(3 \cdot 2)^3$

4. $(3 - 2)^3$

5. $4^2 - 3^2$

6. $8^2 + 15^2$

7. $100(1.1)^2$

8. $200(1.05)^2$

9. $(11 - 4)^2$

10. $(11 + 4)^2$

11. $-13^2 + 4^3$

12. $6^3 + 6^2 + 6 + 1$

13-24. Suppose x = 3 and y = 5. Calculate:

13. xy^2

14. $\frac{x^3 + y^2}{2}$

15. $(\frac{1}{x})^4$

16. $(-y)^2$

17. $9 - y^2$

18. $40 + (-x)^2$

19. $-y^4$

20. $3y^2 - 2y^3$

21. $4x^2 - 2x$

22. $-y^2$

23. $1 + 3(x + y)^2$

24. $(\frac{y - 2}{y + 2})^3$

25-32. Use the formula in this lesson to calculate to the nearest dollar what you would have if you:

25. invest $100 for 2 years at 8%.

C 26. invest $100 for 3 years at 8%.

C 27. invest $500 for 4 years at 6%.

C 28. invest $500 for 5 years at 6%.

C 29. invest $250 for 2 years at 5%.

C 30. invest $250 for 14 years at 5%.

1-2. Each group of simplifications leads to a pattern. Do the simplifications. Then describe the pattern using variables.

1. Simplify: (a) $\dfrac{50^3}{50^2}$ (b) $\dfrac{6^3}{6^2}$ (c) $\dfrac{(\frac{1}{2})^3}{(\frac{1}{2})^2}$ (d) $\dfrac{(-3)^3}{(-3)^2}$

(e) What is the pattern?

2. Simplify: (a) $3^2 \cdot 2^2$ (b) $5^2 \cdot 3^2$ (c) $2^2 \cdot 30^2$

(d) $(-1)^2 \cdot (-5)^2$ (3) What is the pattern?

C 3-12. When a coin has a probability p of landing on heads, the probability of <u>no</u> heads in 10 random tosses is $(1 - p)^{10}$. Calculate this probability to three decimal places if p is:

3. .9   4. .8   5. .7   6. .6   7. .5

8. .4   9. .3   10. .2   11. .1   12. 0

Question for discussion

1. Why is the word "compound" used to describe "compound interest?"

Lesson 4

The Growth Model of Powering

In Chapter 9, you studied <u>linear</u> growth. When powering is used, the growth is called <u>exponential</u>.

<u>Linear growth</u>

1. Begin with amount b.
2. <u>Add</u> m (the slope) for x time periods.
3. Result: b + mx

<u>Exponential growth</u>

1. Begin with amount P.
2. <u>Multiply</u> by B (the base) for x time periods.
3. Result: $P \cdot B^x$

In linear growth, something is added to the original amount. But in exponential growth the original amount is <u>multiplied</u> by some number again and again. You can see the difference in the ways people save money.

<u>Linear</u> growth
(<u>Simple</u> interest)

1. Begin with $100. Deposit $100 at 5% yearly interest. <u>Do not keep</u> interest in account.

2. <u>Add $5</u> each year.
   After 1 year: $100 + $5
   2 years: $100 + $5·2
   3 years: $100 + $5·3
   4 years: $100 + $5·4

   n years: $100 + $5·n

<u>Exponential</u> growth
(<u>Compound</u> interest)

1. Begin with $100. Deposit $100 at 5% yearly interest. <u>Keep</u> interest in account.

2. <u>Multiply by 1.05</u> each year.
   $100(1.05) after 1 year
   $100(1.05)^2 after 2 years
   $100(1.05)^3 after 3 years
   $100(1.05)^4 after 4 years

   $100(1.05)^n after n years

3. Results:
   after 1 year: $105.00
   2 years: $110.00
   3 years: $115.00
   4 years: $120.00
   5 years: $125.00

   10 years: $150.00
   20 years: $200.00

   | The differ-
   ence between
   simple and
   compound
   interest
   widens as
   years go on. |

3. Results:
   $105.00 after 1 year
   $110.25 after 2 years
   $115.76 after 3 years
   $121.75 after 4 years
   $127.62 after 5 years

   $162.88 after 10 years
   $265.32 after 20 years

Whenever the growth of something depends upon the amount present at the time, you should expect exponential growth.

Examples: Exponential growth.

1. <u>Population growth</u>. According to United Nations data, the Earth's population is about 4 billion now (1975) and doubling each 35 years. If this rate would continue:

Years from now	Population (in billions)
0 (now)	$4 = 4$
35	$4 \cdot 2 = 8$
$35 \cdot 2$	$4 \cdot 2^2 = 16$
$35 \cdot 3$	$4 \cdot 2^3 = 32$
$35 \cdot 4$	$4 \cdot 2^4 = 64$
$35n$	$4 \cdot 2^n$

For <u>short</u> periods of time, the bottom line can be used to give a good estimate of the World's population. The number of goods used by the population, like food, fuel, telephones, also may grow exponentially.

2. <u>Glass and light.</u> Suppose a type of glass 1 unit thick allows $\frac{4}{5}$ of light to go through. Then each extra unit allows only $\frac{4}{5}$ to go

467

through.

Thickness	Amount of light through
1 unit	$\frac{4}{5}$ = .80
2 units	$(\frac{4}{5})^2$ = .64
3 units	$(\frac{4}{5})^3$ = .512
⋮	⋮
n units	$(\frac{4}{5})^n$

The same type of relationship often holds for heat. The amount of heat held in or kept out by insulation grows exponentially with the thickness of the insulation. Sound also behaves similarly.

These examples fit the growth model of powering.

Growth model of powering:

> Let B be a positive number. Suppose that a quantity is multiplied by B in every interval of unit length. Then in an interval of length t, the quantity is multiplied by B^t.

Here is how the examples fit the growth model.

	unit length	base B	In length t, quantity is multiplied by
Compound interest:	1 year	1.05	$(1.05)^t$
Population growth:	35 years	2	2^t
Glass and light:	1 unit	$\frac{4}{5}$	$(\frac{4}{5})^t$

In the growth model of powering, the exponent t does not have to be a positive integer. It is possible to have 31.2 years or glass which is $5\frac{1}{2}$ units thick. It is even possible for t to be negative (to represent time in the past).

What happens when t is zero? Read the growth model. In 0 years, the quantity is multiplied by B^0. But in 0 time periods, population, money or light does not change in value. It is as if the quantity were multiplied by 1. For this reason:

> For any positive number B, $B^0 = 1$.

Compare zero multiples and zero powers.

Zero multiple
$0 \cdot m = 0$
Additive identity

Zero power
$B^0 = 1$
Multiplicative identity

The growth model does not work when the base B is negative. We will not consider B^0 when B is not a positive number.

Questions covering the reading

1-10. Tell whether the idea is connected to linear growth or exponential growth.

1. adding   2. multiplying 3. base   4. slope

5. $m \cdot 0$ 6. B^0 7. zero multiple 8. zero power

9. compound interest   10. simple interest

11. Which interest plan gives more money in an account – simple or compound?

12-17. __Compound interest__. If you deposit $500 at 6% yearly interest compounded, what will you have:

12. at the beginning.

13. after 1 year.

14. after 2 years.

15. after 3 years.

16. after n years.

17. after 0 years.

18-21. __Population growth__. According to the data given in this lesson:

18. If you know the world population this year, how do you estimate it 35 years from now?

19. Estimate the population in 2010.

20. Give the formula for the population 35n years after 1975.

21. Suppose n = 0 in Question 20. What happens in the formula?

22-25. __Thickness of glass__. Suppose 1" of a certain glass allows only 75% of light to go through. How much will go through:

22. 2" of this glass.

23. 3" of this glass.

24. 4" of this glass.

25. n" of this glass.

26. What is the growth model for powering?

27. Interpret the growth model when t is zero.

28-33. Simplify.

28. 5^0

29. -2^0

30. $4 \cdot (8 \cdot 10)^0$

31. $4 \cdot x^0$

32. $(4y)^0$

33. $(\frac{1}{2})^0 + (\frac{2}{3})^0$

Questions for exploration

C 1. Inflation. Inflation works like compound interest. An 8% yearly inflation is not uncommon in the world. If you paid $500 for a trip in one country with this kind of inflation, what would you expect to pay:

(a) 1 year later.   (b) 2 years later.

(c) 3 years later.   (d) n years later.

C 2. Credit cards. If you have a credit card, you often must pay a bill within one month or you are charged interest. The maximum allowable interest of 1.5% per month is used by most companies in most states. (Some states do not allow this high an interest.) Suppose you buy an article costing $30.00. At this rate of interest, what would the item cost if you don't pay for:

(a) 1 month.   (b) 2 months.   (c) m months.

(d) 6 months.   (e) a year.

3. Calculate 3^5, 3^4, 3^3, 3^2, 3^1, and 3^0. Extend this pattern to guess at the meaning of 3^{-1}, 3^{-2}, and 3^{-3}.

4. Calculate 5^5, 5^4, 5^3, 5^2, 5^1, and 5^0. Extend this pattern to guess at the meaning of 5^{-1}, 5^{-2}, and 5^{-3}.

C 5. Suppose you deposit $100 at 5% yearly compound interest. In how many years will your money double?

C 6. Repeat Question 5 if you can receive 8% yearly compound interest.

C 7. The Earth's population is about 4 billion and growing at a rate so that it doubles in about 35 years. Then the population P and year Y are connected by the formula

$$P = 4 \cdot 2^{\left(\frac{Y - 1975}{35}\right)}$$

(Everything in the parentheses is an exponent.) By substitution, find the population in (a) 1975, (b) 2010, (c) 2045, and (d) 2080.

1. Banks occasionally advertise plans like the following: "If you deposit $5000 with us, we will send you $25 a month for as long as you keep the money on deposit."

    (a) Approximately what interest rate is the bank paying on the deposit?

    (b) Why do you think the bank would be happy to have you join this plan?

Lesson 5

The Power Property

Two models can tell you what B^5 means.

First model (Repeated multiplication): $B^5 = B \cdot B \cdot B \cdot B \cdot B$

Second model (Growth): B^5 is what a quantity is multiplied by in 5 years if it is multiplied by B every year.

Each model suggests that powers are closely related to multiplication. So you should expect properties which relate multiplication and powers.

Suppose you wish to multiply 2^7 by 2^4. Is there an easy way? Using the repeated multiplication model:

$$2^7 \cdot 2^4 = \underbrace{2 \cdot 2 \ldots \cdot 2}_{7 \text{ factors}} \cdot \underbrace{2 \cdot 2 \cdot 2 \cdot 2}_{4 \text{ factors}}$$

$$= \underbrace{2 \cdot 2 \cdot \ldots \cdot 2}_{11 \text{ factors}}$$

$$= 2^{11}$$

A similar problem involves the growth model. Suppose you save money at 6% yearly interest.

In 3 years you will have $(1.06)^3$ times what you have now.

In 5 years you would have $(1.06)^2$ times what you have in 3 years.

or $(1.06)^2 \cdot (1.06)^3$ what you have now.

But, at 6%, in 5 years you would have $(1.06)^5$ what you have now.

That is, $(1.06)^2 \cdot (1.06)^3 = (1.06)^5$

These examples suggest the <u>power property</u>, a property which we assume true.

For any real numbers for which

B^m and B^n have meaning,

$B^m \cdot B^n = B^{m+n}$.

<u>Power property</u>:

The power property tells you how to multiply two powers with the same base. Expressions with different bases, like

$$2^5 \cdot 3^4$$

cannot usually be simplified.

<u>Questions covering the reading</u>

1. What does x^6 mean:
    (a) using the repeated multiplication model of powering?
    (b) using the growth model of powering?

2. Explain how to simplify $15^3 \cdot 15^7$ using the repeated multiplication model.

3. Explain how to simplify $(1.06)^6 \cdot (1.06)^{11}$ using the growth model.

4. State the power property.

5-16. Simplify:

5. $3^2 \cdot 3^3$

6. $8^5 \cdot 8^0$

7. $(1.12)^7 \cdot (1.12)^7$

8. $x^5 \cdot x^{100}$

9. $(-1)^4 \cdot (-1)^3$

10. $2^0 \cdot 2^1 \cdot 2^2$

11. $2^3 \cdot 3^5$

12. $(-\frac{1}{2})^3 \cdot (-\frac{1}{2})^1$

13. $7^0 \cdot 7^0 \cdot 7^0$

14. $15^4 \cdot 15^4 \cdot 15^5$ 15. $2^x \cdot 2^y$ 16. $(\frac{3}{4})^a \cdot (\frac{3}{4})^3$

Questions extending the reading

1. A 10-question true-false test is given. (a) How many sequences of T's and F's are possible for the first 3 questions? (b) How many sequences of T's and F's are possible for the last 7 questions? (c) How many sequences of T's and F's are possible for the entire test?

2. A certain photograph enlarger can enlarge any picture to $2\frac{1}{2}$ times its original size. By how much will a picture be enlarged if you use the enlarger: (a) 2 times? (b) 3 times? (c) 5 times?

3. Suppose a shade can let in 3/5 of the light which hits it. Then 3 shades can let in _____ of the light. 2 more shades would let in _____ more of the light. Altogether the 5 shades would let in _____ of the light.

4. Suppose that the population of a colony of bacteria triples each 24 hours. Then in 5 days, there will be _____ times what there is now. In 6 more days there will be _____ times what there is in 5 days. So in 11 days there will be _____ what there is now.

5. Water blocks out light. (At a depth of 10 meters it is not as bright as on the surface.) The amount of light blocked depends upon the purity of the water and the plant and animal life in the water. Suppose one meter of water lets in x of the light. (A possible value for x is .85.) How much light would get through 10 meters of depth? How much would get through 11 meters of depth?

6-11. Give the prime factorization of each number. (For example, the prime factorization of 40 is $2^3 \cdot 5$.) Then give the prime factorization of the product of the two given numbers.

6. 40, 80   7. 144, 27   8. 210, 375

9. 50, 50   10. 36, 48   11. 42, 55

12-15. The given numbers are written in scientific notation. Multiply the numbers.

C 12. $3.24 \cdot 10^5$, $1.3 \cdot 10^3$ 13. $2.0 \cdot 10^8$, $3.5 \cdot 10^0$

C 14. $8.6 \cdot 10^{13}$, $2.1 \cdot 10^2$ 15. $6 \cdot 10^{27}$, $3 \cdot 10^9$

16-19. A thousand is 10^3, a million is 10^6, a billion is 10^9. Write each number as a power of 10.

16. a million billion   17. a billion million

18. a thousand million   19. a thousand thousand

20-35. Simplify as much as possible.

20. $ax \cdot ax$ 21. $3y \cdot y$

22. $5m \cdot 6m^{10}$ 23. $4v^3 \cdot 2v^4$

24. $a^2 b \cdot b^2 a$ 25. $3m^2 n^2 \cdot 4n^3 \cdot 2m$

26. $v^0 \cdot v^2 \cdot w^1 \cdot w^3$ 27. $\frac{1}{2} tu \cdot 6u^2$

28. $3x^2 \cdot y^3 \cdot 4y^2 \cdot x^5$ 29. $(5a^5 b)(6a^2 b^2)$

30. $(7s^0 t^2)(\frac{1}{7} s^2 t)(3s)$ 31. $-w^2(4w^3 m)$

32. $5^3 \cdot 6^2 \cdot 3 \cdot 2 \cdot 5^2$ 33. $3k^2 w^2 \cdot k^3 w^3$

34. $x^2 \cdot 5x^3 \cdot y$ 35. $4^9 \cdot 3^7 \cdot 4^2 \cdot 3$

36-41. Apply both the distributive and power properties to simplify.

36. $a^2(a^3 + 4a^4)$ 37. $2x(3x^2 + 5)$ 38. $y^3(9y^2 + 6y + 1)$

39. $a^9(2a - b)$ 40. $x(xy - x^2)$ 41. $12v(8w + 2v^3 - 6v^4)$

42. How can you check problems like 20-41? Pick a problem, do it, and check your answer.

Lesson 6

<u>Negative Exponents</u>

There are many many properties of real numbers. A feature of these properties is that they do not disagree with each other. Two people doing the same things with real numbers will get the same answers even if they use different methods and properties. This feature of mathematics is called <u>consistency</u>.

To show the consistency of mathematics, we now consider the problem:

What does B^{-n} mean?

We answer this question using three different methods:

(1) by a pattern

(2) by the growth model of powering

(3) using the power property

Each way gives the same answer, stated here.

<u>Negative Exponent Theorem</u>:

B^{-n} is the reciprocal of B^n.

<u>Verifying the theorem by a pattern</u>: Consider these powers of 2. The exponents get smaller.

$$2^5 = 32$$
$$2^4 = 16$$
$$2^3 = 8$$
$$2^2 = 4$$

$$2^1 = 2$$
$$2^0 = 1$$

Notice that each number is half the one above it.

Continuing the pattern:

$$2^{-1} = \frac{1}{2}$$
$$2^{-2} = \frac{1}{4}$$
$$2^{-3} = \frac{1}{8}$$
$$2^{-4} = \frac{1}{16}$$
$$2^{-5} = \frac{1}{32}$$

Notice that 2^5 and 2^{-5} are reciprocals... they multiply to 1. The same is true of 2^4 and 2^{-4}. Notice also that all of these powers of 2 are positive numbers.

Verifying the theorem by a growth model: The world's population is increasing about 2% each year. So to find the population for some year, multiply the previous year's estimate by 1.02.

(Population previous year) • 1.02 = Population this year

Solving for the previous year's population,

$$\text{Pop. prev. year} = \frac{\text{Pop. this year}}{1.02}$$

So divide a year's estimate by 1.02 to get the previous year's population. Let us begin with 1975. According to United Nations estimates, the world's population was about 4.01 billion.

Year	Population (in billions)
1977	$4.01 (1.02)^2$
1976	$4.01 (1.02)$
1975	4.01

We now keep going back.

1974	$\dfrac{4.01}{1.02}$	$\approx$ 3.93
1973	$\dfrac{4.01}{(1.02)^2}$	$\approx$ 3.85
1972	$\dfrac{4.01}{(1.02)^3}$	$\approx$ 3.78

There is a general formula relating year and population.

$$1975 + n \qquad 4.01 \cdot (1.02)^n$$

For 1974, n = -1 (one year back). This suggests that

$$4.01 \cdot (1.02)^{-1} \text{ is } \frac{4.01}{1.02}$$

For 1973, n = -2 (two years back). This suggests that

$$4.01 \cdot (1.02)^{-2} \text{ is } \frac{4.01}{(1.02)^2}$$

Continuing this process, we see that

$$(1.02)^{-n} = \frac{1}{(1.02)^n}$$

In short, $(1.02)^{-n}$ is the reciprocal of $(1.02)^n$.

   <u>Verifying the theorem using the power property.</u> According to
the power property,

$$B^n \cdot B^{-n} = B^{(n + -n)}$$
$$= B^0$$
$$= 1$$

Since B^n and B^{-n} multiply to 1, they must be reciprocals.

   For example, to simplify 20^{-2}, think: "20^{-2} is the reciprocal
of 20^2. 20^2 is 400. So 20^{-2} must be $\frac{1}{400}$."

Questions covering the reading

1. In mathematics, what is meant by consistency?

2. Give the next three equations in this pattern:

$$4^3 = 64; \quad 4^2 = 16; \quad 4^1 = 4; \quad 4^0 = 1;$$

3-8. Give the reciprocal of each number.

3. 6

4. $\frac{1}{5}$

5. $\frac{2}{3}$

6. 2^3

7. $(1.06)^{11}$

8. 8^{-7}

9-12. Suppose the world's population in year 1975 + n is approximately $4.01 (1.02)^n$ billion.

9. What value of n gives the population in 1974?

10. Estimate $(1.02)^{-1}$ to 4 decimal places.

11. What value of n gives the population in 1972?

12. What expression gives the estimated population in 1965?

13-24. Write without a negative exponent.

13. 2^{-3}

14. 3^{-1}

15. 8^{-1}

16. 5^{-2}

17. $(\frac{1}{5})^{-3}$

18. $(\frac{1}{3})^{-2}$

19. x^{-4}

20. y^{-2}

21. $(1.5)^{-4}$

22. $(.8)^{-3}$

23. $9^{-2} \cdot 9^{-1}$

24. $6^{-3} \cdot 6^5$

25. What is the Negative Exponent Theorem?

Questions testing understanding of the reading

1. According to the power property, $B^{-5} \cdot B^6 = \underline{\qquad}$.
   Check your answer by substituting 2 for B.

2. Tell why 2^{-115} is not a negative number.

3-5. The Earth's population was about 4 billion in 1975 and doubling every 35 years. (This is very nearly true.) Then the population P and year Y are related by the formula first given on page 471.

$$P = 4 \cdot 2^{\left(\frac{Y - 1975}{35}\right)}$$

Everything in parentheses is an exponent.

3. Find P when Y is 1940. 4. Estimate the population in 1905.

5. What is P when Y is 1800?

C 6-7. Suppose that you have $100 invested at 6% annual interest compounded yearly. Then in n years you will have $100(1.06)^n$.

6. Approximate $100(1.06)^n$ when $n = -1$. What does your answer mean?

7. Approximate $100(1.06)^n$ when $n = -2$. What does your answer mean?

C 8. Give a decimal approximation to 4^n when n is -1, -2, -3, and -4.

9. Give a decimal approximation to 10^n when n is -1, -2, -3, and -4.

10-15. Simplify:

10. $\dfrac{1}{6^{-2}}$

11. $\dfrac{3}{9^{-1}}$

12. $\dfrac{10y}{x^{-5}}$

13. $\dfrac{2^{-4}}{2^{-3}}$

14. $\dfrac{x^0}{x^{-2}}$

15. $\dfrac{(\frac{1}{3})^9}{(\frac{1}{3})^{-11}}$

16-19. Write each number as a power of 2. You may need negative or zero exponents.

16. the number of possible sequences of 7 H's and T's

17. the probability of 7 heads in 7 tosses of a fair coin

18. the probability of guessing at n True-False questions and getting them all correct

19. the number of moons of the planet Earth

Chapter Summary

The operation called powering arises from three types of situations. The first is <u>repeated multiplication</u>: $B^n = \underbrace{B \cdot B \ \ldots \ \cdot B}_{n \text{ factors}}$

The second is the <u>permutation</u> model: The number of sequences possible when each of n blanks can be filled in B ways is B^n. Its most important applications are to probability. The third is the <u>growth</u> model: If a quantity is multiplied by B in every time period of unit length, then it is multiplied by B^t in a time period of length t.

The growth model is the most general and supplies a meaning for both zero and negative exponents. B^0 is how much a quantity is multiplied by in 0 time, so $B^0 = 1$. B^{-t} is how much the quantity changes going back t time periods. So $B^t \cdot B^{-t}$ is what the quantity is multiplied by when you first go forward time t, then back time t. No change occurs. So $B^t \cdot B^{-t} = 1$. This means that B^t and B^{-t} are reciprocals.

The connection of powering with repeated multiplication leads to many relationships between powers and products. The most basic

482

of these is the <u>power property</u>: $B^m \cdot B^n = B^{m+n}$. More properties relating powers and multiplication are studied in the next chapter.

Powering is an important operation because it is basic to many applications. These include counting problems, population growth, sound or light amplifiers, and--most important for the consumer--compound interest.

CHAPTER 11

OPERATIONS WITH POWERS

Lesson 1

Adding Powers

Certain powers of 10 have common names. (The meaning of "billion" and "trillion" is different in some countries other than the U.S.)

$$10^{-6} = \text{one millionth}$$
$$10^{-3} = \text{one thousandth}$$
$$10^{0} = \text{one}$$
$$10^{3} = \text{one thousand}$$
$$10^{6} = \text{one million}$$
$$10^{9} = \text{one billion}$$
$$10^{12} = \text{one trillion}$$

It is possible to pick two of these powers and multiply them.

$$10^{3} \cdot 10^{9} = 10^{12}$$

thousand · billion = trillion

(Numbers this large occur often. The Gross National Product or GNP is the money value of all goods and services produced in a given year. The GNP of the United States first surpassed one thousand billion dollars in 1971. That's over a trillion dollars. Since then it has been higher.)

You can check this multiplication the long way.

$$1000 \cdot 1,000,000,000 = 1,000,000,000,000$$

But what happens if you want to <u>add</u> a thousand to a billion?

$$1000 + 1,000,000,000 = ?$$

The answer is not simple. <u>When powers are added, you should never expect a simple answer.</u>

For instance, it is <u>impossible</u> to simplify even $\underline{x^2 + x^3}$. But you can simplify $x^3 + x^3$. Using the distributive property:

$$x^3 + x^3 = 1 \cdot x^3 + 1 \cdot x^3 = (1 + 1)x^3 = 2x^3$$

Here are other examples.

$$14y^3 + -5y^3 = 9y^3$$

$$6a^5 - 4a^5 - \frac{1}{3}a^5 = (6 - 4 - \frac{1}{3})a^5 = \frac{5}{3}a^5$$

$$50B^3 + 6B^9 + 11B^9 = 50B^3 + 17B^9$$

In general, you can only simplify sums by adding coefficients of the same power of a variable.

<u>Questions covering the reading</u>

1-8. Give the common name for each number.

1. 10^6 2. 10^3 3. 10^{-6} 4. 10^{-3}

5. 10^{12} 6. 10^{-12} 7. 10^9 8. 10^0

9-14. Write as a power of 10.

9. thousand   10. million

11. billion   12. millionth

485

13. thousandth   14. trillion

15. <u>Multiple choice</u>. The GNP of the United States first surpassed
    y dollars in 1971, where y is:

    (a) 10^3 (b) 10^6 (c) 10^9 (d) 10^{12}

16. What is the GNP?

17. Suppose a city with 10^6 people has 10^4 tourists. How many people
    total are in the city?

18. Write as a decimal: $10^7 + 10^2$.

19-32. Simplify if possible.

19. $y^3 + y^5$ 20. $z^7 - z^6$

21. $3x^2 + 2x^2$ 22. $5a^{60} + 7a^{60}$

23. $\frac{1}{2}m^7 - \frac{1}{2}m^7$ 24. $x + x^5 - 3x^6$

25. $4B^3 - 8B^3 + 12B^3$ 26. $9y^4 + 6y^4 - 2y^4$

27. $-v^3 + v^2 + 2v^3$ 28. $m^2 + m^3 + m^4$

29. $2x^2 + 5x^2 - 7x^3$ 30. $-9w^2 - 3w^2$

31. $x^4 + y^3 + z^2$ 32. $-2n^2 + 12n^2 - 3n + 4$

<u>Questions extending the reading</u>

1-4. Write each number as a power of 10.

1. thousand million   2. million million

3. thousand thousand   4. million billion

5-6. As you know, $x^2 + x^3$ cannot be simplified. Given are attempts to simplify. Show that the simplification is wrong by substituting 3 for x.

5. $x^2 + x^3 \overset{?}{=} x^5$

6. $x^2 + x^3 \overset{?}{=} 2x^5$

7-8. The expression $x^2 + y^2$ cannot be simplified. Given are two attempts to simplify. Show that the simplification is wrong by substituting 5 for x and 4 for y.

7. $x^2 + y^2 \overset{?}{=} (xy)^4$

8. $x^2 + y^2 \overset{?}{=} (x + y)^2$

9. In a city with $2 \cdot 10^5$ people, suppose $2 \cdot 10^3$ leave for vacations. How many people are left?

10-11. The nearest star (other than the Sun) is about 25 million milli miles from the Earth.

10. Write this distance as a decimal.

11. Convert this distance to kilometers.

12-14. In the United States, there are names for numbers for every third power of 10 until 10^{63}, one vigintillion. Look in a dictionary to find the meaning of each term.

12. quadrillion   13. quintillion   14. decillion

15-20. Use the distributive property to help simplify.

15. $3x^4 + 2(2x^4 + x^3)$

16. $-9a^6 - 3(a^5 + a^6)$

17. $5y^6 + \frac{1}{2}(4 + 4y^6)$

18. $2(c^5 - c^9) + 4(c^9 + c^5)$

19. $a^2 + b^2 - (a^2 + b^2)$

20. $3m^{462} - (6m^{231} - 2m^{462})$

Lesson 2

Polynomials

The largest money matters most adults deal with are

    payments on loans for cars, trips, etc.,

    insurance,

    social security payments and benefits,

    home mortgages.

Each of them involves paying or receiving money each month or every

few months or every year. But how much total is paid or received?

The answer is not so easy because interest is involved. Here is an

example of this kind of situation.

> Each birthday from age 12 on Mary has received $50
> from her grandparents. She saves the money and
> can get 7% interest a year. How much will she have
> by the time she is 16?

On 12th birthday:
    50 $= 50$

By 13th birthday:
    $50(1.07) + 50$ $= 103.50$

By 14th birthday:
    $50(1.07)^2 + 50(1.07) + 50$ ≈ 160.74

By 15th birthday:
    $50(1.07)^3 + 50(1.07)^2 + 50(1.07) + 50$ ≈ 222.00

By 16th birthday:
    $50(1.07)^4 + 50(1.07)^3 + 50(1.07)^2 + 50(1.07) + 50 \approx 287.53$

        $\uparrow$ $\uparrow$ $\uparrow$ $\uparrow$ $\nwarrow$

from 12th   from 13th   from 14th   from 15th from 16th
birthday   birthday   birthday   birthday   birthday

The total of $287.53 is $37.53 more than the $250 she received as gifts.

In general, if the scale factor is x, then at age 16 Mary would have

$$50x^4 + 50x^3 + 50x^2 + 50x + 50$$

Suppose other relatives gave Mary 20, 25, 60, 40, and 45 dollars on these 5 birthdays. Then she would have this much more:

$$20x^4 + 25x^3 + 60x^2 + 40x + 45$$

The total she has received and saved is

$$(50x^4 + 50x^3 + 50x^2 + 50x + 50) + (20x^4 + 25x^3 + 60x^2 + 40x + 45)$$

By the assemblage property and using distributivity

$$= (50 + 20)x^4 + (50 + 25)x^3 + (50 + 60)x^2 + (50 + 40)x + 50 + 45$$

$$= 70x^4 + 75x^3 + 110x^2 + 90x + 95$$

Notice that Mary received a total of $70 on her 12th birthday. This money has 4 years to grow. The $75 from her 13th birthday has 3 years to grow. And so on. So the final answer could have been obtained from the given information.

The above expressions are <u>polynomials</u> in the variable x. The largest exponent is the <u>degree</u> of the polynomial.

Examples: Polynomials

1. $70x^4 + 75x^3 + 110x^2 + 90x + 95$ is a polynomial in x of degree 4. 70 is the coefficient of x^4, 75 is the coefficient of x^3, and so on.

2. $4.6y^{10} + 8y^{23} - y$ is a polynomial in y of degree 23. -1 is the coefficient of y. 0 is the coefficient of y^{20}—that's why you don't see y^{20}. The coefficient of y^{10} is 4.6.

In general, a polynomial of degree n in the variable x has the form

$$a_0 + a_1 x + a_2 x^2 + a_3 x^3 + \ldots + a_n x^n$$

The numbers $a_0, a_1, \ldots, a_n$ are the coefficients of the polynomial.

Notice that nothing is done to shorten the expression $70x^4 + 75x^3 + 110x^2 + 90x + 95$. This is because it <u>cannot</u> be shortened. Polynomials can only be simplified when they contain the same powers of the same variable.

<u>Questions covering the reading</u>

1-5. Huey, Dewey, and Louie are triplets. They received the following cash presents on their birthdays.

	<u>H</u>	<u>D</u>	<u>L</u>
in 1972:	$10	$15	$5
in 1973:	$20	$15	$40
in 1974:	$15	$15	nothing

They put their money into a bank account which pays 6% yearly interest.

1.   How much did Huey have by his 1973 birthday?

2.   How much did Dewey have by his 1973 birthday?

3.   How much did Louie have by his 1973 birthday?

4.   How much did Huey have by his 1974 birthday?

5.   Who had more by his 1974 birthday, Dewey or Louie?

6-9. Use the amounts of Questions 1-5. In 1975, Huey received $30 on his birthday. If he had invested all money into an account with scale factor x, how much would he have had by his birthday:

6. in 1972   7. in 1973   8. in 1974   9. in 1975

10. Give an example of a polynomial in x of degree 4.

11. Give an example of a polynomial in y of degree 5.

12. Give an example of a polynomial in z of degree 2.

13-16. For the polynomial $2m^3 - 5m^4 + m - 80$, give the coefficient of:

13. m^4 14. m^3 15. m^2 16. m

17-20. For the polynomial $3 + 4x^2 - 90x^3 - x^5$, give the coefficient of:

17. x^2 18. x 19. x^3 20. x^5

21. Give the degree of the polynomial of Questions 13-16.

22. Give the degree of the polynomial of Questions 17-20.

23-26. Simplify each expression as much as possible.

23. $(2x^3 + 8x^2 - 9x + 6) + (10x^3 - 8x^2 - 2x - 12)$

24. $1 + m + m^2 + m^3 + m^4$

25. $-3 + 5y^2 - 7y^2 + 9y^2 - 11y^2 + 13$

26. $2x^3 - x^4 + 2x + \frac{1}{2} - 3.1x + 3x^4$

27. Give two examples of money matters which involve paying or receiving money periodically.

Questions testing understanding of the reading

1-4. Give the value of $x^3 + 2x^2 - 9x + 2$ when x is:

1. 1   2. 2   3. -1   4. 0

491

5-8. Give the value of $8m^3 - 30m$ when m is:

5. 3 6. $\frac{1}{2}$ 7. 2 8. -1

9-14. Suppose a person pays \$500 a year for life insurance. The insurance company can multiply the amount by m each year. How much will the insurance company have after:

9. 1 year (include 2nd year's payment)

10. 2 years 11. 3 years 12. 4 years

C   13. Give the answer to Question 11 if m is 1.10.

C   14. Give the answer to Question 12 if m is 1.10.

15-18. After 5 years of birthdays, Wanda has received and saved

$$80x^4 + 60x^3 + 70x^2 + 45x + 22.93$$

dollars, having invested the money at a scale factor of x/year.

15. How much did Wanda get on her most recent birthday?

16. How much did Wanda get on the first year's birthday?

17. What is a reasonable value for x in this problem?

18. If $x \doteq 1$, how much has Wanda saved? What does the value of 1 of x mean?

19-23. Persons A, B, and C, and D have the following amounts of money:

A: $100y^5 + 2y^4 + 80y^3 + 45y + 30$

B: $90y^3 + 20y^2 + 60y + 100$

C: $8y^4 + 2y^3 + 8y^2 + 7.50y + 12$

D: $50y^2 + 100y + 500$

19. A and B together have _____.

20. C and D together have _____.

C   21. If $y = 1.08$, calculate how much B has. Note: $(1.08)^3 \approx 1.26$.

492

C   22. If $y = 1.08$, calculate how much D has.

23. If 2 other people have the same amount of money as Person A, how much do the three people have all together?

24-27. Simplify:

24. $3(4x^2 - 12x + 6) - 2(8 - 4x^2 + 3)$

25. $9t^3 + 9t^2 + 9t + 9 - (8t^2 + 8t + 8)$

26. $-14v^4 - 8v^3 + 12v^2 - 6v + 11 - (v^4 + 10v^3 + 6v^2 - 8v + 12)$

27. $w^2 - w^5 - w^3 + w^4 + 1 - 2w^3 - w^2 - w$

28. How can you check your answer to Question 24?

29-30. Below are four instances of a pattern.

$$1^2 = 1$$
$$1^2 + 2^2 = 5$$
$$1^2 + 2^2 + 3^2 = 14$$
$$1^2 + 2^2 + 3^2 + 4^2 = 30$$

The general pattern may be described using variables.

$$1^2 + 2^2 + \ldots + n^2 = \frac{1}{3}n^3 + \frac{1}{2}n^2 + \frac{1}{6}n$$

At right is a formula for the sum of the squares of the integers from 1 to n.

29. Use the formula to find the sum of the squares of the integers from 1 to 10. Check your answer by actual addition of the squares.

30. Use the formula to find the sum of the squares of the integers from 1 to 40.

Lesson 3

Dividing Powers

Adding powers leads to polynomials. Since subtractions can be converted to additions, subtracting powers also leads to polynomials.

But multiplying powers of the same number gives another power of that number. $x^m \cdot x^n = x^{m+n}$ Divisions can be converted to multiplications. So you should expect dividing powers of the same number to result in a power of that number. This is exactly what happens.

<table>
<tr><td>Division of Powers
Theorem:</td><td>For any real numbers for which B^m and B^n are defined,

$$\frac{B^m}{B^n} = B^{m-n}$$</td></tr>
</table>

Why is this so? Using properties you have had:

$$\frac{B^m}{B^n} = B^m \cdot \frac{1}{B^n} \qquad \text{(Def. of division)}$$

$$= B^m \cdot B^{-n} \qquad \text{(Negative Exponent Theorem)}$$

$$= B^{m + -n} \qquad \text{(Power Property)}$$

$$= B^{m-n} \qquad \text{(Def. of subtraction)}$$

Examples: Dividing Powers

1. $\dfrac{2^{10}}{2^3} = 2^{10-3} = 2^7$ This is easy to check by calculating 2^{10} and 2^3 and doing the division.

494

2. $\dfrac{x^7}{x^3} = x^{7-3} = x^4$ 3. $\dfrac{x^3}{x^7} = x^{3-7} = x^{-4}$

Some people prefer the answer $\dfrac{1}{x^4}$ to Example 3 because that answer has no negative exponents.

4. You would expect $\dfrac{y^9}{y^9}$ to equal 1. By the Division of Powers theorem,

$$\frac{y^9}{y^9} = y^{9-9} = y^0 = 1.$$

5. $\dfrac{6z^{10}}{18z} = \dfrac{6}{18} \cdot \dfrac{z^{10}}{z^1} = \dfrac{1}{3} \cdot z^{10-1} = \dfrac{1}{3}z^9$ (It all can be done in one step.)

6. $\dfrac{3a^2 b}{\frac{1}{2}b^5} = \dfrac{3}{\frac{1}{2}} \cdot a^2 \cdot \dfrac{b^1}{b^5} = 6 \cdot a^2 \cdot b^{-4}$

   $= 6a^2 b^{-4}$ ⟵——— Some people like this answer. It has no fractions.

   $= \dfrac{6a^2}{b^4}$ ⟵——— Some people like this answer better. It has no negative exponents.

There are about 215 million people in the United States. Some McDonald's restaurants advertise that over 12 billion hamburgers have been sold by all restaurants in the chain. How many hamburgers is that per person in the U.S.?

The question asks for a rate. So division is required.

no. of hamburgers per person $= \dfrac{\text{no. of hamburgers}}{\text{no. of people}} = \dfrac{12 \text{ billion}}{215 \text{ million}}$

Even with a calculator, dividing numbers in the billions by numbers in the millions is not easy. Most calculators don't have enough space for all the numbers. The best way to do this division is to write billion and million as powers of 10.

$$\frac{12 \text{ billion}}{215 \text{ million}} = \frac{12 \cdot 10^9}{215 \cdot 10^6} = \frac{12}{215} \cdot \frac{10^9}{10^6}$$

$$\approx .0558 \cdot 10^3$$

$$= 55.8 \text{ hamburgers}$$

The Division of Powers Theorem only applies when the bases are the same. When the bases are different, nothing much can be done.

$$\frac{13^5}{5^4} \text{ cannot be easily simplified.}$$

Questions covering the reading

1. What is the Division of Powers Theorem?

2. Give reasons for each step.

(a) $\dfrac{2^{40}}{2^{30}} = 2^{40} \cdot \dfrac{1}{2^{30}}$

(b) $\phantom{\dfrac{2^{40}}{2^{30}}} = 2^{40} \cdot 2^{-30}$

(c) $\phantom{\dfrac{2^{40}}{2^{30}}} = 2^{40 + -30}$

$\phantom{\dfrac{2^{40}}{2^{30}}} = 2^{10}$

3-26. (a) Write without fractions. (b) If your answer to part (a) has negative exponents, rewrite the answer so that there are fractions but no negative exponents.

3. $\dfrac{y^{10}}{y^6}$

4. $\dfrac{y^6}{y^{10}}$

5. $\dfrac{z^{15}}{2z^5}$

6. $\dfrac{x^2}{x}$

7. $\dfrac{2z^5}{z^{15}}$

8. $\dfrac{v^9}{v^9}$

9. $\dfrac{B^2}{B^2 \cdot B^3}$

10. $\dfrac{60d^{60}}{3d^3}$

11. $\dfrac{a^5 \cdot a^3}{a^{-2}}$

12. $\dfrac{5m^{-2}}{20m^6}$

13. $\dfrac{\text{one million}}{\text{one billion}}$

14. $\dfrac{\text{five thousand}}{\text{one thousandth}}$

15. $\dfrac{\text{one billionth}}{\text{ten billion}}$

16. $\dfrac{6^3 \cdot 5^4}{6^8 \cdot 5^2}$

17. $\dfrac{2^{15}}{3^5}$

18. $\dfrac{3.2 \cdot 10^5}{1.6 \cdot 10^2}$

19. $\dfrac{\frac{1}{2}v^9}{\frac{1}{3}v^5}$

20. $\dfrac{\frac{3}{5}x^6}{\frac{2}{3}x^{-7}}$

21. $\dfrac{2^8}{3^6}$

22. $\dfrac{z^{-5}}{z^{-3}}$

23. $\dfrac{200a^{-1}}{450a^{-4}}$

24. $x^{-9} \div x^{-3}$

25. $x^{-9} \cdot x^{-3}$

26. $\dfrac{3mn}{6n^2}$

27-29. Use a population of 210 million for the United States.

27. In 1972, General Motors had sales of about 30 billion dollars. How many dollars is that per person in the United States?

28. On January 1, 1971, there were 7 hundred TV stations in the United States (including possessions). If everyone was tuned into one of these, about how many people would be viewing each station?

C 29. The U.S. National Debt is the amount that the government has borrowed, usually from taxpayers or businesses in the form of bonds (like savings bonds). In 1976, the debt was around 400 billion dollars. How much is the debt per person in the United States?

497

Questions applying the reading

1. Light travels about 186 thousand miles per second. How long does it take light to travel from the Sun to the Earth, about 93 million miles away?

C 2. The human population of the Earth is about 4.1 billion. The land area of the Earth is about 146 million sq km. How many people is this per sq km?

3-6. One micron = 10^{-6} meter. How many microns are in one:

3. meter   4. kilometer   5. centimeter   6. millimeter

7. One milligram = 10^{-3} grams. One kilogram = 10^3 grams. Weights of vitamins are given in milligrams. Weights of people are often given in kilograms. How many milligrams are in one kilogram?

8. How many millimeters are in one kilometer? (Hint: Will Question 7 help?)

9-12. Use the prime factorizations to help simplify the fractions.

$$1620 = 2^2 \cdot 3^4 \cdot 5 \qquad\qquad 1296 = 2^4 \cdot 3^4$$

$$693 = 3^2 \cdot 7 \cdot 11 \qquad\qquad 432 = 2^4 \cdot 3^3$$

9. $\dfrac{1620}{693}$ 10. $\dfrac{432}{1620}$ 11. $\dfrac{432}{1296}$ 12. $\dfrac{1620}{1296}$

13-20. Simplify these rather complicated expressions.

13. $\dfrac{a^3 b^2 c^4}{abc}$

14. $\dfrac{2^9 x^2 y^4}{2^7 x^3 y^3}$

15. $\dfrac{(t + u)^5}{(t + u)^4}$

16. $\dfrac{3(a - 2b)^4}{(a - 2b)^{12}}$

17. $\dfrac{r^2}{s^3} \cdot \dfrac{s^2}{r^3}$

18. $\dfrac{4a^3}{9b} \cdot \dfrac{3b^4}{8a}$

19. $\dfrac{\dfrac{x^2}{y^2}}{\dfrac{x^3}{y^5}}$

20. $\dfrac{\dfrac{3mn^2}{2m^2n}}{\dfrac{5n^3}{12m^4}}$

Lesson 4

Powers of Powers

A wall is designed to deaden sound. Suppose 1 mm thickness of the wall multiplies sound intensity by B. Then 10 mm or 1 cm thickness will multiply sound intensity by B^{10}. By how much will 3cm thickness multiply sound intensity?

Answer 1: Think in centimeters. 1 cm thickness multiplies by B^{10}. So 3 cm thickness multiplies by $(B^{10})^3$.

Answer 2: Think in millimeters. 3 cm = 30 mm. 1 mm thickness multiplies by B. So 30 mm thickness multiplies by B^{30}.

The answers must be equal. So $(B^{10})^3 = B^{30}$.

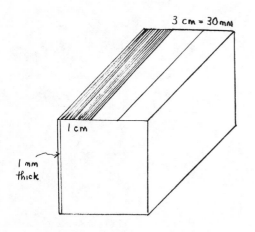

This argument could be repeated with any base B and thickness. It demonstrates:

Power of a Power Theorem:

$$(B^m)^n = B^{mn}$$

The Power of a Power Theorem can be verified with the repeated multiplication model.

1. $(x^2)^3 = x^2 \cdot x^2 \cdot x^2$ by the repeated multiplication model of powering

   $= x^{2+2+2}$ by the power property

   $= x^6$

2. The theorem predicts $(5^3)^{-4} = 5^{-12}$. This can be verified.

$$(5^3)^{-4} = \frac{1}{(5^3)^4} = \frac{1}{5^3 \cdot 5^3 \cdot 5^3 \cdot 5^3} = \frac{1}{5^{3+3+3+3}} = \frac{1}{5^{12}} = 5^{-12}$$

(It's easier just to apply the theorem. $(5^3)^{-4} = 5^{-12}$.)

If the world's population doubles every 35 years in the future, then in 58 $\cdot$ 35 years or only 2030 years from now the population will be

$$4 \cdot 2^{58} \text{ or } 2^2 \cdot 2^{58} \text{ or } 2^{60}.$$

This is about equal to the number of square inches on the Earth! Clearly the population cannot grow at its present rate even only for 2000 more years. But how big is 2^{60}? You can use the Power of a Power Theorem to help approximate 2^{60}.

Earlier we have calculated 2^{10}.

$$2^{10} = 1024 \approx 1000 = 10^3$$

That is, $\qquad 2^{10} \approx 10^3$

Then $\qquad (2^{10})^6 \approx (10^3)^6$

So $\qquad 2^{60} \approx 10^{18}$

10^{18} can be done in your head. It is a 1 followed by 18 0's. Thus

$$2^{60} \approx 1,000,000,000,000,000,000.$$

1-8. Simplify:

1. $(2^3)^2$ 	2. $(5^2)^{-1}$ 	3. $(x^8)^2$ 	4. $3(y^{-3})^{-2}$

5. $((\frac{1}{2})^0)^{-1}$ 	6. $((\frac{2}{3})^{-2})^{-2}$ 	7. $-3(a^2)^4$ 	8. $(B^5)^{-5}$

9-12. $2^{10} \approx 10^3$. Use this information to approximate:

9. 2^{20} 	10. 2^{-10} 	11. 2^{140} 	12. 2^{30}

13-15. Use the repeated multiplication model to verify that:

13. $(x^3)^5 = x^{15}$ 	14. $((36)^{-2})^4 = 36^{-8}$

15. $(2^{15})^3 = 2^{45}$

16-18. Suppose a sheet of $\frac{1}{4}$" thickness will let only $\frac{2}{3}$ of sound go through. Then how will a sheet of the given thickness affect sound?

16. 1 inch 	17. 1 foot 	18. 1 yard

1-12. Simplify:

1. $((3^5)^4)^3$ 	2. $3 \cdot (2^1)^2$

3. $(x^{-1})^{-1} + 3x$ 	4. $(y^{-2})^2 + (y^{-3})^3$

5. $2(v^2)^3 - 9(v^3)^2$ 	6. $6(w^8)^3 - 6(w^7)^3$

7. $(((((2)^1)^0)^{-1})^{-2})^{-3}$ 	8. $3(x^3)^6 + 2(x^3)^5$

9. $9(m^2)^9 - 9(m^3)^6$ 	10. $5(y^{-1})^{-3} + 5(z^{-1})^{-3}$

11. $(a^0)^3 \cdot (a^1)^2$ 	12. $2B^4 \cdot 3(B^2)^2$

13-14. $\pi^2 \approx 10$. Use this information to approximate:

13. π^{10} 	14. π^6

15-18. You have been told two things about the world's present rate of population growth.

      1. It is growing 2% each year. (That is, it is being multiplied by 1.02 each year.)

      2. It is doubling every 35 years.

These mean about the same thing because in 35 years of 2% growth, the population is multiplied by $(1.02)^{35}$, and $(1.02)^{35} \approx 2$. Use this information to help answer the following questions. What will the population be multiplied by in:

15. 70 years.
           16. 350 years.

17. 140 years.
         18. 35n years.

19-21. $(1.07)^6 \approx 1.5$. Use this information to approximate:

19. $(1.07)^{12}$
      20. $30(1.07)^6$
      21. $(1.07)^{-6}$

22-25. Which number is not equal to the other two?

22. 3^{20}, 6^{15}, 9^{10}
          23. 12^2, 4^6, 2^{12}

24. $(x^2)^3$, $(x^3)^2$, $x^3 \cdot x^2$
      25. m^7, $(\frac{1}{m})^{-7}$, $m^3 + m^4$

26-27. If money is invested at 6% annual interest, it will be multiplied by $(1.06)^{10}$ at the end of a decade. How much will it be multiplied by

26. in 2 decades
      27. in 3 decades

28-30. $(1.08)^9 \approx 2$. Use this information to approximate:

28. $(1.08)^{90}$
      29. $(1.08)^{27}$
      30. $(1.08)^{45}$

31. Suppose you think you can gain 8% a year investing in the stock market. At this rate, if you begin with $500, about how much would you have in 18 years? Use the approximation of Questions 28-30 to make the arithmetic easy.

503

Lesson 5

Powers of Products and Quotients

You have now studied five operations with powers: addition, subtraction, multiplication, division, and powering. Now we begin looking at applying powering after another operation has been done.

The expression $(2t)^5$ is an example of the power of a product. It can be rewritten as follows:

$$(2t)^5 = 2t \cdot 2t \cdot 2t \cdot 2t \cdot 2t$$
$$= 2^5 t^5$$
$$= 32t^5$$

This idea can be used to calculate a power of a number written in scientific notation. (This notation is discussed in detail in the next lesson.)

$$(3.4 \cdot 10^8)^3$$

This is easy using the repeated multiplication model.

$$(3.4 \cdot 10^8)^3 = (3.4 \cdot 10^8) \cdot (3.4 \cdot 10^8) \cdot (3.4 \cdot 10^8)$$

All numbers are multiplied so the assemblage property can be applied.

$$= (3.4)^3 \cdot (10^8)^3$$
$$= (3.4)^3 \cdot 10^{24}$$

When you want a positive integer power of a product, you can always use the repeated multiplication model.

$$(xy)^n = \underbrace{(xy) \cdot (xy) \cdot \ldots \cdot (xy)}_{\text{n factors}}$$

$$= \underbrace{x \cdot x \cdot \ldots \cdot x}_{\text{n factors}} \cdot \underbrace{y \cdot y \cdot \ldots \cdot y}_{\text{n factors}}$$

$$= x^n \cdot y^n$$

This result also holds when n is a negative integer or zero. (See the Questions for discussion.) So there is a simple theorem which helps to calculate the power of a product.

<u>Power of a Product Theorem:</u>

> If n is any integer,
>
> $$(xy)^n = x^n y^n$$

In Chapter 12 you will see that this theorem holds even when n is a fraction.

Examples: Using the Power of a Product Theorem

1. $(5t)^{-3} = 5^{-3} t^{-3} = \dfrac{1}{125} \cdot \dfrac{1}{t^3} = \dfrac{1}{125t^3}$

2.   Suppose you triple the radius of a circle. What happens to the area? (This question has applications to pizza-making.)

   Solution: Let the original radius be k. Then the new radius is 3k. Recall the formula $A = \pi r^2$. Compare the areas of the two circles by dividing.

$$\frac{\text{Area of new circle}}{\text{Area of original circle}} = \frac{\pi(3k)^2}{\pi k^2} = \frac{\pi 3^2 \cdot k^2}{\pi \cdot k^2}$$

$$= 3^2 = 9$$

The new area is 9 times the old.

There is a Power of a Quotient Theorem which is quite similar to the Power of a Product Theorem. You are asked to give reasons for this theorem in the Questions for discussion.

Power of a Quotient Theorem:

> If n is any integer,
>
> $$\left(\frac{x}{y}\right)^n = \frac{x^n}{y^n}$$

Examples:

1. $\left(\frac{2}{3}\right)^5 = \frac{2^5}{3^5} = \frac{32}{243}$. You can verify this by calculating

$\frac{2}{3} \cdot \frac{2}{3} \cdot \frac{2}{3} \cdot \frac{2}{3} \cdot \frac{2}{3}$.

2. $3\left(\frac{x}{2y}\right)^4 = 3 \cdot \frac{x^4}{(2y)^4} = 3 \cdot \frac{x^4}{2^4 y^4} = \frac{3x^4}{16y^4}$.

Notice that the power is calculated before multiplying by 3.

Questions covering the reading

1-21. Rewrite each expression without using parentheses.

1. $(3x)^2$ 2. $\left(\frac{2}{7}z\right)^3$ 3. $(-4y)^3$

4. $(\frac{a}{5})^2$

5. $(\frac{b}{3})^5$

6. $(-\frac{c}{2})^4$

7. $(2v^2)^3$

8. $(-3w^2)^2$

9. $(\frac{x^3}{4})^5$

10. $2(3y)^4$

11. $\frac{1}{2}(2x)^5$

12. $5(4a)^2$

13. $3(-6m)^2$

14. $-9(-\frac{x}{2})^4$

15. $100(\frac{2}{5}w)^3$

16. $(1.2 \cdot 10^3)^2$

17. $(9 \cdot 10^{-4})^2$

18. $(2.3 \cdot 10^8)^3$

19. $(2m)^{-3}$

20. $(\frac{1}{z})^{-4}$

21. $2(6xy)^{-2}$

22. $4(8v)^{-1}$

23. $(\frac{w}{2})^{-1}$

24. $(\frac{2v}{3})^{-3}$

25. $(9x)^m$

26. $(-4y)^n$

27. $(\frac{u}{3})^t$

28-30. What happens to the area of a circle if its radius is:

28. multiplied by 4.   29. divided by 3.   30. multiplied by 10.

31. A 14" pizza is how many times bigger than a 7" pizza with the same thickness?

32. A 15" pizza has how many times more ingredients than a 5" pizza with the same ingredients?

507

1-3. Suppose you are sitting at a distance d from a stereo speaker. The loudness of the sound satisfies the formula $L = \dfrac{k}{d^2}$.

(k is a number which depends on the units you are using. It does not change.) How many times louder is the sound if you move:

1. 3 times as far away from the speaker.

2. half as far from the speaker.

3. $1\frac{1}{2}$ times as far away from the speaker.

4-6. The volume v of a sphere with diameter d satisfies $v = \dfrac{\pi}{6} d^3$. The sun, planets, and moons in the solar system are nearly spheres.

4. Jupiter's diameter is about 11 times the Earth's. So its volume is about _____ times the Earth's volume.

5. The sun's diameter is about 110 times the Earth's. So its volume is about _____ times the Earth's volume.

6. The moon's diameter is about $\frac{1}{4}$ the Earth's. So _____.

7-9. The surface area of a sphere with diameter d is πd^2.

7. If one hollow plastic ball has three times the diameter of a second, how do the areas compare?

8. The moon's diameter is about $\frac{1}{4}$ the Earth's. How do the surface areas compare?

9. The diameter of Mars is about half that of Earth. (a) How does its surface area compare? (b) Since the Earth's surface is about 29% land, does Mars or the Earth have more land area?

10-12. Assume the pizzas have the same thickness.

10. A 10" pizza is how many times bigger than a 7" pizza?

11. A 12" pizza is how many times bigger than a 10" pizza?

12. A 14" pizza is how many times bigger than a 12" pizza?

13-16. True or False.

13. $2^4 \cdot 3^4 = 6^4$ 14. $x^a \cdot y^a = (xy)^a$

15. $\dfrac{m^3}{n^3} = (\dfrac{m}{n})^3$ 16. $3^{50} + 6^{50} = 9^{50}$

17-26. Simplify:

17. $(2x)^2 \cdot (3x^2)^3$ 18. $(\dfrac{m}{5})^3 \cdot (\dfrac{5}{m})^2$

19. $(\dfrac{3t^2}{2})^4 \cdot t^2$ 20. $(9(1.06)^2)^3 \cdot (1.06)$

21. $2(4x)^3 + 5(4x)$ 22. $(4y)^2 \cdot (4y)^{-2}$

23. $4m^{12} - 2(3m^3)^4$ 24. $(\tfrac{1}{2}n)^3 + (\tfrac{3}{2}n)^3$

25. $(2x^2y^3)^6 - (4x^4y^5)^3$ 26. $(ab^2)^3 - (a^2b)^3 + (a^3b)^2 - (ab^3)^2$

27-30. According to <u>Introductory Mathematics for Technicians</u>, by Auerbach and Groza, the power output P of an aircraft propeller is given by the formula

$$P = kr^3D^5$$

where D is the diameter of the propeller and r is the number of revolutions per minute. (k is a number which depends on the units used. It does not change.) How many times more power has plane 1 than plane 2 if:

27. the propeller of plane 1 has $1\tfrac{1}{2}$ times the diameter of plane 2 and they revolve the same number of times per minute?

28. How much is power increased if the number of revolutions of a propeller is doubled?

29. How much more power does plane 3 have if its propeller is $\tfrac{2}{3}$ the diameter of plane 2 but rotates $2\tfrac{1}{2}$ times as fast?

30. How much more power does plane 5 have if its propeller is $\tfrac{3}{4}$ the diameter of plane 4 but rotates twice as fast?

31-33. Write as a power of a number in parentheses.

31. $9v^2w^2$

32. $\dfrac{v^3}{64}$

33. $\dfrac{a^3b^3}{c^6}$

Questions for discussion

1. Give reasons for each step. This shows that the Power of a Product Theorem holds for negative exponents.

   (a) $(xy)^{-n} = \dfrac{1}{(xy)^n}$

   (b) $\qquad = \dfrac{1}{x^n y^n}$

   (c) $\qquad = \dfrac{1}{x^n} \cdot \dfrac{1}{y^n}$

   (d) $\qquad = x^{-n} y^{-n}$

2. Why is it true that $(xy)^0 = x^0 y^0$?

3. Give reasons for each step. These steps demonstrate the Power of a Quotient Theorem.

   (a) $\left(\dfrac{x}{y}\right)^n = (x \cdot \dfrac{1}{y})^n$

   (b) $\qquad = (x \cdot y^{-1})^n$

   (c) $\qquad = x^n \cdot (y^{-1})^n$

   (d) $\qquad = x^n \cdot y^{-n}$

   (e) $\qquad = \dfrac{x^n}{y^n}$

Scientific Notation

From ancient times, people have tried to write numbers in as little space as possible. For example, 26 is short for

~~IIII~~ ~~IIII~~ ~~IIII~~ ~~IIII~~ ~~IIII~~ I

The usual way of writing integers, <u>base 10</u>, is particularly efficient. It allows large and small numbers to be written in a short space. Each digit stands for a multiple of a power of 10.

base 10 notation	expanded base 10 notation
82961 is short for	$8 \cdot 10^4 + 2 \cdot 10^3 + 9 \cdot 10^2 + 6 \cdot 10^1 + 1 \cdot 10^0$
46.7581 is short for	$4 \cdot 10^1 + 6 \cdot 10^0 + 7 \cdot 10^{-1} + 5 \cdot 10^{-2} + 8 \cdot 10^{-3} + 1 \cdot 10^{-4}$
.00002109 is short for	$2 \cdot 10^{-5} + 1 \cdot 10^{-6} + 9 \cdot 10^{-8}$

But sometimes even base 10 notation is not short enough. If you own a calculator, you know that numbers with more than 8 digits cannot be displayed. For these numbers, <u>scientific notation</u> is helpful. This notation was first used by Descartes.

Here are some examples of scientific notation.

Base 10 notation	Scientific notation
4,160,000,000	$4.16 \cdot 10^9$
12	$1.2 \cdot 10^1$
.00346	$3.46 \cdot 10^{-3}$
.0000000006	$6.0 \cdot 10^{-10}$

Definition: | A number is in <u>scientific</u> <u>notation</u> if it is written in the form

$$r \cdot 10^n$$

where r is a real number with $1 \leq r < 10$ and n is an integer.

It is easy to convert from base 10 to scientific notation and back. The power of 10 used is always the first term in expanded base 10 notation. Here are more examples.

<u>Base 10 notation</u>	<u>Scientific notation</u>
-3800	$-3.8 \cdot 10^3$
.0000800003	$8.00003 \cdot 10^{-5}$
7.23	$7.23 \cdot 10^0$

Because it is easy to multiply, divide, or take powers of powers, scientific notation is often easier to work with than base 10 notation.

Questions covering the reading

1-6. Write each number in expanded base 10 notation and scientific notation.

1. 4300

2. 763,000

3. .00000328

4. .051

5. 9.91

6. 26.3

7-10. Write each number in base 10 notation and scientific notation.
Use only two decimal places in the scientific notation.

7. $2 \cdot 10^7 + 6 \cdot 10^6 + 3 \cdot 10^5 + 1 \cdot 10^3$

8. $9 \cdot 10^{-2} + 7 \cdot 10^{-3} + 8 \cdot 10^{-4}$

9. $1 \cdot 10^4 + 3 \cdot 10^2 + 5 \cdot 10^0 + 7 \cdot 10^{-2} + 9 \cdot 10^{-4}$

10. $6 \cdot 10^9 + 4 \cdot 10^7 + 3 \cdot 10^1 + 8 \cdot 10^{-3}$

11-20. Write each number in base 10 notation.

11. $6.670 \cdot 10^{-8}$, the attraction (in dynes) between two objects of
1 gram mass each 1 cm apart.

12. Planck's constant, $4.14 \cdot 10^{-15}$ electron volt-seconds.

13. Avogadro's number, $6.0247 \cdot 10^{23}$ molecules per mole.

14. The mass of a hydrogen atom, about $1.6734 \cdot 10^{-24}$ grams.

15. The mass of the sun, a rather typical star, about $2 \cdot 10^{38}$ grams.

16. $2.99792 \cdot 10^5$ km per second, the approximate speed of light in a
vacuum.

17. That computer can do an arithmetic problem in $2.4 \cdot 10^{-9}$ seconds.

18. The mass of an electron is about $9.1085 \cdot 10^{-28}$ grams.

19. The speed of light in a vacuum is about $3 \cdot 10^{10}$ cm per second.

20. The total number of hands possible in bridge is about $6.35 \cdot 10^{11}$.

21. Give two reasons why scientific notation is helpful.

C 1-14. Use the following numbers:

$$a = 2 \cdot 10^4 \qquad m = 6.83 \cdot 10^{-2} \qquad x = -8 \cdot 10^{14}$$

$$b = 3.2 \cdot 10^6 \qquad n = -2.49 \cdot 10^{-6} \qquad y = 4.1 \cdot 10^{-14}$$

Put into scientific notation.

1. a^3

2. b^2

3. $\dfrac{x}{a}$

4. $\dfrac{x}{b}$

5. $5am$

6. $.25bn$

7. xy

8. $(ax)^2$

9. $(\dfrac{x}{a^2})^5$

10. abx

11. $a + b$

12. $m + n$

13. $3a^2$

14. $-64x^{-2}$

15-16. You can use scientific notation to calculate the number of digits in the base 10 representation of an integer.

15. $2^{10} \approx 10^3$. How many digits are in 2^{100}?

16. $3^{10} \approx 6 \cdot 10^4$. How many digits are in 3^{20}?

17-22. The exponent n in 10^n is a <u>superscript</u>. Computers and calculators cannot write superscripts, so a variant of scientific notation must be used. Some machines write 1.732 E 06 for $1.732 \cdot 10^6$. On these machines, how would the given number from Questions 1-14 be represented?

17. a

18. m

19. x

20. b

21. n

22. y

23-24. Cells in living tissue are of nearly the same size, $3 \cdot 10^{-4}$ cm in length, about half that in width and thickness.

23. What is the volume of a typical cell?

24. How many cells would be in a cubic <u>mm</u> of volume? (Hint: There are 10^3 cubic mm in a cubic cm.)

The connection of powering with repeated multiplication means that relationships with powers and products should be expected. Two such relationships are basic:

Power property (Product of powers): $B^m \cdot B^n = B^{m+n}$

Power of a product: $(xy)^n = x^n y^n$

There are corresponding properties with division.

Quotient of powers: $\dfrac{B^m}{B^n} = B^{m-n}$

Power of a quotient: $\left(\dfrac{x}{y}\right)^n = \dfrac{x^n}{y^n}$

A fifth property is important.

Power of a power: $(B^m)^n = B^{mn}$

Adding powers of the same number x leads to <u>polynomials</u>. A polynomial in x form has the form

$$a_n x^n + a_{n-1} x^{n-1} + \ldots + a_2 x^2 + a_1 x + a_0$$

where the numbers of the form a_i are called <u>coefficients</u> and remain the same throughout a problem. Many investment and savings plans involve polynomial expressions and, as you shall see in later chapters, polynomials arise in other situations as well.

The common way of writing numbers is called base 10 because it is based on powers of 10. Scientific notation also uses powers of 10 and is particularly helpful when working with very large or very small numbers.

CHAPTER 12

SQUARES AND SQUARE ROOTS

Lesson 1

<u>The Definition of Square Root</u>

The customary definition of square root is in terms of repeated multiplication. Since $3 \cdot 3 = 9$, 3 is called a square root of 9. Since $-5 \cdot -5 = 25$, -5 is a square root of 25. In general:

<u>Definition of</u>
<u>Square Root:</u>

Suppose $x \cdot x = A$. Then x is a <u>square root</u> of A.

Examples: Square roots.

1. $-3 \cdot -3 = 9$. So -3 is a square root of 9.

2. $12m^3 \cdot 12m^3 = 144m^6$. So $12m^3$ is a square root of $144m^6$.

Example 1 shows that 9 has two square roots, 3 and -3. The symbol $\sqrt{}$ stands only for the <u>positive</u> square root.

$$\sqrt{9} = 3$$

The negative square root of 9 is written $-\sqrt{9}$.

$$-\sqrt{9} = -3$$

From the definition of square root,

$$\underline{\sqrt{A} \cdot \sqrt{A} = A,}$$

This is the fundamental property linking square roots and repeated multiplication. For example, since $\sqrt{9}$ = 3,

$$\sqrt{9} \cdot \sqrt{9} = 9.$$

But it is also true that

$$\sqrt{8} \cdot \sqrt{8} = 8.$$

But $\sqrt{8}$ is not an integer. It can only be approximated by a decimal. To get an approximation, pick a number and multiply it by itself. (That is, square the number.)

$$2.8 \cdot 2.8 = 7.24$$

This tells us that $2.8 = \sqrt{7.24}$. So 2.8 is too small to be $\sqrt{8}$.

$$2.9 \cdot 2.9 = 8.41$$

This says 2.9 is larger than $\sqrt{8}$. As a result

$$2.8 < \sqrt{8} < 2.9$$

You could write $\qquad \sqrt{8} = 2.8...$

To get an approximation correct to two decimal places, square the numbers 2.81, 2.82, 2.83, etc.

$$2.82 \cdot 2.82 = 7.9524 \qquad \text{So } 2.82 < \sqrt{8}$$
$$2.83 \cdot 2.83 = 8.0089 \qquad \text{So } 2.83 > \sqrt{8}$$

This indicates that $2.82 < \sqrt{8} < 2.83$. You could write $\sqrt{8} = 2.82...$

Of course, a calculator with a $\sqrt{}$ key can get you many decimal places quickly. The author's calculator shows

$$\sqrt{8} = 2.8284271\ldots$$

On pages 742 and 743 are approximations to the square roots of all integers from 1 to 100.

Square roots have a great number of applications. In this chapter, three major types of applications are discussed.

distance	(Lessons 2 and 8)
growth	(Lesson 4)
statistics	(Lesson 9)

There are many other applications. For example, an application used by police is on page 522, Questions 21-23.

Questions covering the reading

1. Give the two square roots of 9.

2. Give the two square roots of 36.

3. The symbol for the positive square root of 9 is _____.

4. In words, $-\sqrt{19}$ means_____.

5. The two square roots of $\frac{1}{2}$ are written _____ and _____.

6-9. Simplify:

6. $\sqrt{49}$ 7. $-\sqrt{49}$ 8. $-\sqrt{121}$ 9. $\sqrt{400}$

10. What is the definition of square root?

11-14. Simplify:

11. $\sqrt{16} \cdot \sqrt{16}$ 12. $\sqrt{x} \cdot \sqrt{x}$ 13. $\sqrt{12} \cdot \sqrt{12}$ 14. $2\sqrt{3} \cdot 5\sqrt{3}$

15. Since $-4m \cdot -4m = 16m^2$, _____ is a square root of _____.

16. Since $2^{26} = (2^{13})^2$, _____ is a square root of _____.

$$7^2 = 49$$
$$7.1^2 = 50.41$$
$$7.2^2 = 51.84$$
$$7.3^2 = 53.29$$
$$7.4^2 = 54.76$$
$$7.5^2 = 56.25$$

17-20. According to the above information:

17. $\sqrt{50}$ is between _____ and _____.

18. $\sqrt{55}$ is between _____ and _____.

19. $\sqrt{54.531}$ is between _____ and _____.

20. 7.3 is a square root of _____.

21-24. Between what two integers does the positive square root of each number lie?

21. 40   22. 72   23. 100.324   24. 1.5

25. What major types of applications of square roots are found in this chapter?

Questions testing understanding of the reading

C 1-4. Approximate the positive square root of each number correct to one decimal place. If you have a calculator with a $\sqrt{\ }$ key, check the calculation by squaring the approximation.

1. 104   2. 450   3. 300   4. 3005

5. $(\frac{1}{2})^2 = \frac{1}{4}$. So _____ is a square root of _____.

6-9. Name the two square roots of each number.

6. 4900   7. x^2 8. $\frac{4}{9}$ 9. $\frac{1}{17}$

521

10. If x is a square root of y, then $x^2 = $ _____.

11. If r is a square root of 0, then _____.

12. Name all integers between 1 and 99 whose square roots are integers.

13. Name all integers between 99 and 200 whose square roots are integers.

14-20. Simplify.

14. $\sqrt{(-3)^2}$ 15. $\sqrt{2^2 + 3^2}$ 16. $\sqrt{2^2} + \sqrt{3^2}$ 17. $\sqrt{\dfrac{16}{81}}$

18. $\sqrt{-4 \cdot -4}$ 19. $-\sqrt{3} \cdot \sqrt{3}$ 20. $\sqrt{2} \cdot \sqrt{2} \cdot \sqrt{3} \cdot \sqrt{3} \cdot \sqrt{4}$

21-23. <u>How fast was that car going?</u> Suppose that the 4 tires of a car skid for L feet before stopping. Then let r be the speed of the car. Police officers sometimes estimate r by using the following formulas.

On a dry concrete road, $r \approx \sqrt{24L}$

On a wet concrete road, $r \approx \sqrt{12L}$

Estimate the minimum speed of a car (to the nearest mph) if:

21. skid marks are 30 feet long on a dry concrete road.

22. skid marks are 30 feet long on a wet concrete road.

23. skid marks are 42 feet long on a dry concrete road.

Lesson 2

Square Roots and Lengths

Where does the name "square root" come from? It may surprise you to learn that "square roots" are related both to squares in geometry and roots of plants.

A plant rests on its roots. To the ancient Greeks, a square was thought of as an area. It rested on a side called its <u>square root</u>.

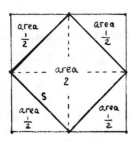

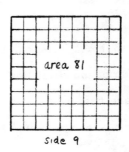

side 9

In the example drawn above, the side has length 9. The area is $9 \cdot 9$ or 81. As you know, $\sqrt{81} = 9$.

The big square has area 4. The area of the tilted square is 2. If s is the length of a side of the tilted square, then $s \cdot s = 2$. So $s = \sqrt{2}$.

From these examples comes the area property of square roots:
If a square has area A, then the length of each side is $\sqrt{A}$. For
instance, the tilted square below has area 10. So the length of a
side is $\sqrt{10}$.

Area of large square = 16

Area of each triangle = 1.5

So area of tilted square = 10

In decimals, $\sqrt{10}$ = 3.16... You can check this value by measuring
the length of the side of the tilted square.

Many lengths are best described by using square roots. For this
reason, square roots are often found in measuring distances and in
geometry.

The area A of a square with side x satisfies the formula

$$A = x^2$$

When the area A is 25, it is clear that x must be 5. If x stands
for a length, the replacement set for x is the set of positive real
numbers. So there is no other answer. But suppose you only are
given only the equation

$$25 = x^2$$

524

Then it is possible that x could be -5. $(-5)^2 = 25$. So there are two solutions, -5 and 5. The solution set is $\{5, -5\}$.

<u>Questions covering the reading</u>

1. How is a "square root" related to a square?

2. If the area of a square is 12, then the length of a side of that square is _____.

3. Must all sides of a square be the same length?

4-5. Find (a) the area and (b) the length of a side of the tilted square. (Hint: Get the area of the large square first, then cut off the right triangles.)

4.

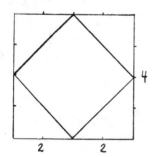

5.

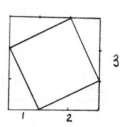

6-9. The area of a square satisfies the formula $A = s^2$. Find s when:

6. $A = 49$ 　　　　　　7. $A = 8$

8. $A = 12$ sq cm 　　　　9. $A = 90000$ sq m

10-15. Solve each equation. Remember that negative numbers might be solutions.

10. $x^2 = 16$ 　　　11. $1 = y^2$ 　　　12. $z^2 = 0$

13. $6 = a^2$ 　　　14. $b^2 = 2.25$ 　　15. $11 = c^2$

Questions extending the reading

1-4. Solve each equation.

1. $2x^2 = 20$

2. $\frac{1}{3} = 3y^2$

3. $z^2 - 3 = 2$

4. $w^2 + 46 = 91$

5. Given that a and b are positive, solve for t: $at^2 = b$.

6-7. Find the length of a side of the tilted square by using Questions 4-5, page 525.   Then make a large copy of each drawing with each unit equal to 1 cm. Measure the length of a side of the tilted square to check your answer.

6.

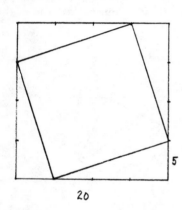

20

7.

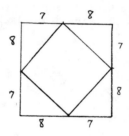

Question for discussion

1. The sign $\sqrt{\ }$ is called a "radical" sign, and some people say "radical 2" for $\sqrt{2}$. (Most say "square root of 2. ") What plant has a name very similar to "radical?" (The name of this plant and the word "radical" both come from the Latin word <u>radix</u>, meaning "root. ")

Lesson 3

The Exponent $\frac{1}{2}$

Situation 1: Think of a colony of bacteria whose population multiplied by 9 in 48 hours. Then in 24 hours at this rate, the population was multiplied by 3.

Situation 2: If a window pane 1 cm thick lets in $\frac{81}{100}$ of light, then a $\frac{1}{2}$ cm pane of the same glass will let in $\frac{9}{10}$ of light.

These situations point out a relationship between square roots and growth.

Theorem:

> If a quantity is multiplied by B in a given time period, then in $\frac{1}{2}$ the time (at the same rate) the quantity is multiplied by $\sqrt{B}$.

How can this be verified in general? Look at the following diagram.

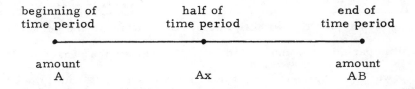

beginning of time period	half of time period	end of time period
amount A	Ax	amount AB

The amount A is multiplied by B in the entire time period. It is

527

multiplied by x in the first half of the period. If the rate is constant, it will be multiplied by x in the second half also. Then

$$(Ax)x = AB$$
$$Ax^2 = AB$$
$$x^2 = B$$

Since the scale factor x is positive in this application, $x = \sqrt{B}$.

For example, suppose money earns 10% yearly interest accumulated daily. It will then double in about 7 years. At this rate, what happens in 3 1/2 years?

beginning	$3\frac{1}{2}$ years later	7 years later

$100	$100x	$2 \cdot \$100$

$$(100x)x = 2 \cdot 100$$
$$100x^2 = 200$$
$$x^2 = 2$$
$$x = \sqrt{2} \text{ because } x \text{ is positive}$$

The $100 is multiplied by $\sqrt{2}$, yielding about $141.42 in 3 1/2 years.

When a quantity is multiplied by B in a given time period, in n time periods it is multiplied by B^n. (This is the growth model of powering.) So in 1/2 of a time period it should be multiplied by $B^{\frac{1}{2}}$. But the above situation shows it is multiplied by $\sqrt{B}$. Thus $B^{\frac{1}{2}}$ and $\sqrt{B}$ must be the same.

<u>Square Root-Power</u>
<u>Theorem</u>:

$$B^{\frac{1}{2}} = \sqrt{B}$$

Examples: $\frac{1}{2}$ powers.

1. $9^{\frac{1}{2}} = \sqrt{9} = 3$ 2. $(\frac{9}{4})^{\frac{1}{2}} = \sqrt{\frac{9}{4}} = \frac{3}{2}$

3. $-6 \cdot 2^{\frac{1}{2}} = -6\sqrt{2} \approx -8.48$

The Square Root-Power Theorem is helpful because it allows all properties of powers to be applied to square roots. For instance, to calculate the square root of a power

$$\sqrt{x^{20}}$$

translate to powers

$$= (x^{20})^{\frac{1}{2}}$$

and now use the Power of a Power Theorem $(x^m)^n = x^{mn}$.

$$= x^{10}$$

There is one tricky thing which happens with square roots of variables. The $\sqrt{}$ sign stands for the <u>positive</u> square root. If you get an answer which could be negative, use the $|\ |$ absolute value sign to make sure it is positive.

529

$$x^6 = (x^6)^{\frac{1}{2}} = \left| x^3 \right|$$

When x is negative, x^3 is negative, so the $|\ |$ sign is necessary.

Caution: $B^{\frac{1}{2}}$ and $\frac{1}{2}B$ are not at all the same.

$\frac{1}{2}B + \frac{1}{2}B = B$ When you **add** $\frac{1}{2}B$ to itself, you get B.

$B^{\frac{1}{2}} \cdot B^{\frac{1}{2}} = B$ When you **multiply** $B^{\frac{1}{2}}$ by itself, you get B.

In the expression B^t, t may be **any** number. You now have seen what happens if $t = \frac{1}{2}$. For this course, we do not need to consider other values of t. You will study other rational exponents (as in $2^{\frac{3}{8}}$) and even irrational exponents (as in $2^{\sqrt{5}}$) in your next algebra course.

Questions covering the reading

1-2. Pick your answer from the following choices.

(a) $B^{\frac{1}{2}}$ (b) B^2 (c) $2B$ (d) $\frac{1}{2}B$

1. If the population of a country is multiplied by B each year, then in 2 years it is multiplied by _____.

2. If the population of a country is multiplied by B each year, then in 1/2 year it is multiplied by _____.

3-6. Suppose a pane of glass 20mm thick will let in 4/9 of the light. How much light will be let in by a pane of the same glass whose thickness is:

3. 40mm   4. 60mm   5. 10mm   6. 50mm

7. Use the growth model of powering to explain what $B^{\frac{1}{2}}$ means.

8. Give an example to show that $x^{\frac{1}{2}}$ and $\frac{1}{2}x$ are different.

9-16. Write without an exponent.

9. $9^{\frac{1}{2}}$

10. $-9^{\frac{1}{2}}$

11. $8^{\frac{1}{2}}$

12. $(\frac{2}{3})^{\frac{1}{2}}$

13. $(\frac{16}{9})^{\frac{1}{2}}$

14. $100^{\frac{1}{2}}$

15. $7^{\frac{1}{2}} \cdot 7^{\frac{1}{2}}$

16. $5 \cdot (\frac{1}{3})^{\frac{1}{2}} \cdot (\frac{1}{3})^{\frac{1}{2}}$

17-22. Given that x = 16 and y = 25, calculate

17. $x^{\frac{1}{2}} \cdot y^{\frac{1}{2}}$

18. $x^{\frac{1}{2}} + y^{\frac{1}{2}}$

19. $-(xy)^{\frac{1}{2}}$

20. $(x + y)^{\frac{1}{2}}$

21. $16x^{\frac{1}{2}} - 3y^{\frac{1}{2}}$

22. $(x^2 + y^2)^{\frac{1}{2}}$

23. When a = -2, a^2 = _____.

24. Choose the correct answer. For any real number x, $\sqrt{x^2}$ =

(a) x   (b) -x   (c) $|x|$

25-30. Simplify.

25. $\sqrt{x^8}$

26. $\sqrt{10^{10}}$

27. $\sqrt{y^4}$

28. $\sqrt{a^6}$

29. $\sqrt{B^{12}}$

30. $\sqrt{c^6}$

Questions extending the reading

1. In 100 days, a cancerous tumor grew 9 times in size. After 50 days, it was probably about _____ times larger than at the start.

2. If money earns 5% yearly interest, it will double in value in about 14 years. What will happen in 7 years?

3. If money is invested at 5.5% yearly interest, it will triple in about 20 years. What will happen in 10 years?

4-6. Given that x is a positive number, simplify:

4. $(4x^{16})^{\frac{1}{2}}$

5. $4(\frac{16}{9})^{\frac{1}{2}}$

6. $9(x^2)^{\frac{1}{2}}$

7-9. Simplify:

7. $\left(6^{\frac{1}{2}}\right)^6$

8. $\left(9^{\frac{1}{2}} + 9^{\frac{1}{2}}\right)^2$

9. $\left(x^{\frac{1}{2}}\right)^{10} + \left(y^{\frac{1}{2}}\right)^4$

10. Recall the formula estimating the world population P in the year Y.

$$P = 4 \cdot 2^{\frac{Y - 1975}{35}}$$

According to this formula, estimate the population to the nearest hundred million in:

(a) 1975

(b) 2010

(c) $1992\frac{1}{2}$ (the middle of 1992)

11. Find a value of x for which $\sqrt{x^{10}}$ does not equal x^5.

Questions for discussion

1-7. It is possible to deal with $\frac{3}{2}, \frac{5}{2}, \frac{7}{2}, \ldots$, as exponents in the following way:

$$49^{\frac{3}{2}} = 49^{1 + \frac{1}{2}} = 49^1 \cdot 49^{\frac{1}{2}} = 49 \cdot 7 = 343$$

$$3^{\frac{5}{2}} = 3^2 \cdot 3^{\frac{1}{2}} = 9\sqrt{3} \approx 15.5$$

Give a decimal approximation to each number. Use the tables of square roots to help.

1. $4^{\frac{3}{2}}$ 2. $9^{\frac{3}{2}}$ 3. $2^{\frac{5}{2}}$

4. $2^{\frac{7}{2}}$ 5. $10^{\frac{3}{2}}$ 6. $100^{\frac{7}{2}}$

7. 49^0, $49^{\frac{1}{2}}$, 49^1, $49^{1\frac{1}{2}}$, 49^2, $49^{2\frac{1}{2}}$, and 49^3.

8. (Use the formula of the previous Question 10 and the ideas of Questions 1-7.) Estimate to the nearest hundred million the world population in:

   (a) $2027\frac{1}{2}$ (b) 2045   (c) the middle of 2062

Lesson 4

The Middle of Growth

Graphed below are solutions to the equation $y = 2^x$.

x	y
-2	.25
-1	.5
0	1
1	2
2	4
3	8

Graphed below is the population of California every 10 years from 1890 to 1960.

Year	Population
1890	1,213,398
1900	1,485,053
1910	2,377,549
1920	3,426,861
1930	5,677,251
1940	6,907,387
1950	10,586,223
1960	15,717,204

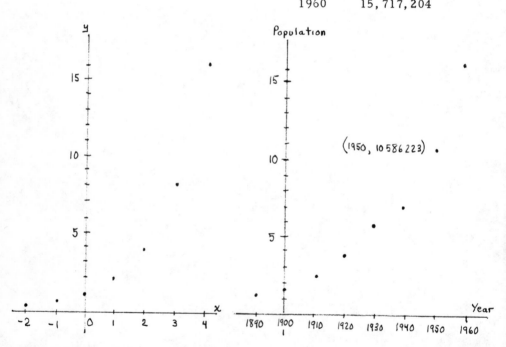

The graphs are very similar. At left the y values double every time the exponent x is increased by 1. This is _exponential_ growth.

Over short periods of time population sometimes grows exponentially. The graph shows that the population of California was nearly exponential for the years 1890 - 1960. This is an unusually long time for such growth.

To find out how fast California grew, divide to compare population sizes each 10 years. This gives a scale factor for the growth each decade. For the two decades from 1940 to 1960:

$$\frac{\text{population of California in 1950}}{\text{population of California in 1940}} = \frac{10,586,223}{6,907,387} \approx 1.53$$

$$\frac{\text{population of California in 1960}}{\text{population of California in 1950}} = \frac{15,717,204}{10,586,223} \approx 1.48$$

Over the 70 year period, the scale factor is between 1.21 and 1.65 per decade. (This is a rate of growth of between 21% and 65% each decade.) Because many decades show nearly the same growth, we say the population has grown exponentially.

Now we ask: What was the population in 1915? All we can do is estimate. Here is one way to estimate. <u>Assume the growth was at the same rate for the years 1910-1920.</u> Then each 5 years will have the same growth factor.

1. growth factor from 1915 to 1920 = growth factor from 1910 to 1915.

2. The growth factor is found by dividing.

$$\frac{\text{population in 1920}}{\text{population in 1915}} = \frac{\text{population in 1915}}{\text{population in 1910}}$$

We want to know the population in 1915. Call this P. Then, filling in data,

3. $$\frac{3,426,861}{P} = \frac{P}{2,377,549}$$

This is a proportion. Using the Means-Extremes Property you learned in Chapter 6,

4. $$P^2 = 2,377,549 \cdot 3,426,861$$

By the definition of square root, since P is positive:

5. $$P = \sqrt{2,377,549 \cdot 3,426,861}$$

6. Using a calculator: $\qquad P \approx 2,860,000$

Compare steps 3 and 5. In step 3, the equation has form

$$\frac{P}{a} = \frac{b}{P}$$

a and b being the 1910 and 1920 populations. In step 5,

$$P = \sqrt{ab}$$

This process can be repeated with any type of growth problem.

Middle of Growth Theorem:	Consider a quantity which is changing exponentially. Suppose there is a of this quantity at the beginning of a time period and there is b at the end. Then in the middle of this period there is $\sqrt{ab}$.

Examples: Using the Middle of Growth Theorem

1. Suppose the population P of a country was 4 million in 1920 and 9 million in 1940. Then, if P grew exponentially, in 1930 (the middle of the time period)

$$P = \sqrt{4 \cdot 9} = \sqrt{36} = 6$$

To check compare the growth factors.

$$\frac{\text{Pop. in 1930}}{\text{Pop. in 1920}} = \frac{6}{4} = 1.5 \qquad \frac{\text{Pop. in 1940}}{\text{Pop. in 1930}} = \frac{9}{6} = 1.5$$

They are equal so the answer checks.

2. If a bank compounds interest "continuously," your money grows at an even rate during a year. If you begin a year with $100 and end it with $107.50, then in the middle you would have

$$\sqrt{100 \cdot 107.50}$$
$$= \sqrt{10750}$$
$$\approx \$103.71$$

When a and b are positive, $\sqrt{ab}$ is always between a and b. It is called the <u>geometric mean</u> of a and b. The above examples show:

The geometric mean of 4 and 9 is $\sqrt{4 \cdot 9}$ or 6.

The geometric mean of 100 and 107.50 is exactly $\sqrt{10750}$ or about 103.71.

<u>Questions covering the reading</u>

1-10. Give all solutions to each equation.

1. $x^2 = 4$ 2. $9 = y^2$ 3. $z^2 = 8$

4. $\dfrac{P^2}{3} = 12$ 5. $Q^2 = 1.44$ 6. $t^2 = 14400$

7. $\dfrac{3}{x} = \dfrac{x}{5}$ 8. $\dfrac{y}{-2} = \dfrac{-8}{y}$ 9. $\dfrac{27P}{8} = \dfrac{3}{2P}$

10. $\dfrac{n}{-4} = \dfrac{-9}{n}$

11. Over the years 1890-1960, California had a growth factor of about 1.44 per decade. What does the 1.44 mean?

C 12. The population of West Virginia was 1,860,000 in 1960 and 1,744,000 in 1970. What was its growth factor for the decade 1960-1970?

13. What equation can be solved to estimate California's population in 1915?

14-22. Calculate the geometric mean of the given numbers correct to 1 decimal place.

14. 3 and 4   15. 6 and 8   16. 1 and 9

17. 2 and 32   18. $\dfrac{3}{2}$ and $\dfrac{4}{3}$ 19. 9 and 16

20. 1 and 3   21. 1.3 and 5.2   22. 10 and 5

23-25. Give the geometric mean of the given positive numbers.

23. a and b   24. x and 4y   25. 1 and P

26. What is the Middle of Growth Theorem?

27. According to some estimates, the Earth's population will be 7 billion in the year 2000. If it was 4 billion in 1976, estimate the population in 1988.

28. If you invest $1 at 9% a year compound interest, you will have $2 in about 8 years. How much would you have in 4 years?

Questions applying the reading

C 1. In 1970 the population of California was 19,953,134. (a) Using the 1960 population what would a growth factor of 1.44 per decade have predicted? (b) Would this have been a good prediction?

C 2. In 1880 the population of California was 864,694. (a) Using the 1890 population, what would a growth factor of 1.44 per decade have predicted? (b) Is this an accurate prediction?

C 3-5. Estimate the population of California (to the nearest hundred thousand) in

3. 1925   4. 1935   5. 1955

6. If a right angled piece of stone is chopped off to form a right triangle with sides a and b as pictured, then the height of the stone is $\sqrt{ab}$. How high is the stone if a = 3 and b = 7? Calculate to the nearest tenth.

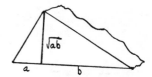

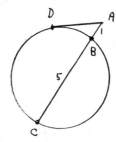

7. The length of $\overline{AD}$ is the geometric mean of the lengths of AB and AC. With AB = 1 and $\overline{BC}$ = 5, what is the length of $\overline{AD}$?

8-10. These exercises are related.

8. Give the geometric mean of x^3 and x^5.

9. Give the geometric mean of t^{10} and t^{20}.

10. Generalize the results of Questions 8 and 9.

1. Here are steps in going from $\dfrac{P}{a} = \dfrac{b}{P}$ to $P = \sqrt{ab}$. What was done in each step?

$$\text{Given} \quad \dfrac{P}{a} = \dfrac{b}{P}$$

(a) $$P^2 = ab$$

(b) $$P = \sqrt{ab} \quad \text{or} \quad P = -\sqrt{ab}$$

(c) Why is $-\sqrt{ab}$ often not chosen as a solution?

12

2

2. Suppose a square has the same area as the rectangle at left. (a) What is the length of a side of the square? (b) Generalize part (a).

Lesson 5

Simplifying $\sqrt{xy}$

The expression $\sqrt{xy}$ often occurs from questions of growth. So it is nice to know that $\sqrt{xy}$ is often easy to simplify.

Recall the Power of a Product Theorem:

$$(xy)^n = x^n \cdot y^n$$

When $n = \frac{1}{2}$: $\qquad\qquad (xy)^{\frac{1}{2}} = x^{\frac{1}{2}} \cdot y^{\frac{1}{2}}$

Now we use the $\sqrt{}$ symbol instead of the $\frac{1}{2}$ power.

Square Root-Product Theorem:	If x and y are positive real numbers, then $\sqrt{xy} = \sqrt{x} \cdot \sqrt{y}$.

The theorem does not look like it simplifies anything. But watch how it is used.

Examples: Applying the Square Root-Product Theorem.

1. To simplify $\sqrt{200}$, first factor 200 so that one factor is a perfect square.

$$\sqrt{200} = \sqrt{100 \cdot 2}$$

Apply the theorem. $\qquad = \sqrt{100} \cdot \sqrt{2}$

Now it's easy. $\qquad\qquad = 10\sqrt{2}$

So $\sqrt{200} = 10\sqrt{2}$. Since $\sqrt{2} \approx 1.414$ (from tables), a decimal approximation to $\sqrt{200}$ is $10(1.414)$ or 14.14.

2. Another example of the same type. 9 is a perfect square factor of 45.

$$\sqrt{45} = \sqrt{9 \cdot 5}$$
$$= \sqrt{9} \cdot \sqrt{5}$$
$$= 3\sqrt{5}$$

Look in the tables on page 742 to verify that this is correct.

3. On page 534, the estimated population of California in 1915 was calculated from $\sqrt{2,377,549 \cdot 3,426,861}$. This arithmetic is hard even with calculators. To do the arithmetic without calculators, first estimate. (This is a realistic thing to do anyway.)

$$\sqrt{2,377,549 \cdot 3,426,861} \approx \sqrt{2,380,000 \cdot 3,430,000}$$

Scientific notation: $= \sqrt{2.38 \cdot 10^6 \cdot 3.43 \cdot 10^6}$

$$\approx \sqrt{8.16 \cdot 10^{12}}$$

Applying the theorem: $= \sqrt{8.16} \cdot \sqrt{10^{12}}$

$$\approx 2.86 \cdot 10^6 = 2,860,000$$

In all of these examples, the idea is the same. Factor the number inside the $\sqrt{\ }$ sign. Look for factors which are perfect squares. Then apply the theorem. The perfect square factors will have nice square roots and simplify.

Recall that there is a Power of a Quotient Theorem. (See page 506 if you have forgotten.)

$$\left(\frac{x}{y}\right)^n = \frac{x^n}{y^n}$$

So $(\frac{x}{y})^{\frac{1}{2}} = \dfrac{x^{\frac{1}{2}}}{y^{\frac{1}{2}}}$

Translating to the $\sqrt{\ }$ sign yields a theorem about square roots.

Square Root-Quotient Theorem:

If x and y are positive real numbers, then

$$\sqrt{\frac{x}{y}} = \frac{\sqrt{x}}{\sqrt{y}}$$

For example, $\sqrt{\dfrac{7}{4}} = \dfrac{\sqrt{7}}{\sqrt{4}} = \dfrac{\sqrt{7}}{2}$

$$\approx \frac{2.646}{2} = 1.323$$

Caution: There are no simple properties relating square roots and addition. For example, the sum of $\sqrt{9}$ and $\sqrt{25}$ is 8, but there is no obvious relationship between 9, 25, and 8.

Questions covering the reading

1-2. Two of the three numbers are equal. Which two?

1. $\sqrt{15}$, $\sqrt{3} + \sqrt{5}$, $\sqrt{3} \cdot \sqrt{5}$ 2. $2\sqrt{3}$, $3\sqrt{2}$, $\sqrt{12}$

3-14. Simplify. Do not approximate by decimals.

3. $\sqrt{18}$ 4. $\sqrt{1800}$ 5. $\sqrt{40}$

6. $\sqrt{60}$ 7. $\sqrt{8}$ 8. $\sqrt{27}$

9. $\sqrt{4 \cdot 10^6}$ 10. $\sqrt{9,000,000}$ 11. $\sqrt{2 \cdot 10^4}$

12. $\sqrt{75}$ 13. $\sqrt{250}$ 14. $\sqrt{125}$

15-24. Calculators are not allowed for these exercises. Use the tables on pages 742 and 743 to help estimate these square roots to the nearest hundredth.

15. $\sqrt{300}$ 16. $\sqrt{325}$

17. $\sqrt{338}$ 18. $\sqrt{2000}$

19. $\sqrt{2000 \cdot 18000}$ 20. $\sqrt{1,432,000 \cdot 1,960,000}$

21. $\sqrt{\dfrac{2}{9}}$ 22. $\sqrt{\dfrac{12}{100}}$

23. $\sqrt{\dfrac{5}{16}}$ 24. $\sqrt{\dfrac{48}{49}}$

25-30. True or false.

25. $\dfrac{\sqrt{100}}{\sqrt{25}} = 4$ 26. $\sqrt{9} \cdot \sqrt{4} = \sqrt{36}$

27. $\sqrt{3^2} \cdot \sqrt{2^2} = 6$ 28. $\sqrt{3^2 + 2^2} = 5$

29. $\sqrt{\dfrac{3}{4}} = \dfrac{\sqrt{3}}{2}$ 30. $\sqrt{x^2 + y^2} = x + y$

Questions extending the reading

1-4. The Power of a Product Theorem can be used to simplify square roots involving variables. For example,

$$\sqrt{100x^8} = \sqrt{100} \cdot \sqrt{x^8} = 10x^4$$

Assume all variables are positive. Simplify.

1. $\sqrt{x^2 y^2}$ 2. $\sqrt{9y^4}$

3. $\sqrt{100a^4 b^2}$ 4. $\sqrt{4y^{14}}$

5. $\sqrt{\dfrac{4}{x^2}}$ 6. $\sqrt{\dfrac{10}{z^{10}}}$

7. If the variables in Questions 1-6 were not positive, how would the answers be affected?

8-9. Given that $\sqrt{153} \approx 12.37$, estimate:

8. $\sqrt{15300}$ 9. $\sqrt{1530000}$

10-11. Given that $\sqrt{85} \approx 9.22$, estimate:

10. $\sqrt{.85}$ 11. $\sqrt{.00000085}$

12-17. Simplify using the distributive property.

12. $3\sqrt{2} + 5\sqrt{2}$ 13. $9\sqrt{3} - 11\sqrt{3}$

14. $\dfrac{\sqrt{5}}{2} + \dfrac{\sqrt{5}}{2}$ 15. $\sqrt{x} + 3\sqrt{x}$

16. $6\sqrt{y} - \dfrac{1}{2}\sqrt{y}$ 17. $\sqrt{z} + \sqrt{4z} + \sqrt{9z}$

18. (a) Simplify $\sqrt{12}$, $\sqrt{27}$, and $\sqrt{75}$.

    (b) True or false: $\sqrt{12} + \sqrt{27} = \sqrt{75}$.

19. Repeat Question 18 for $\sqrt{27}$, $\sqrt{63}$, and $\sqrt{90}$.

20. Repeat Question 18 for $\sqrt{150}$, $\sqrt{6}$, and $\sqrt{216}$.

Questions for discussion

1-5. These questions are related.

1. Suppose a number is multiplied by 4. What happens to its square root? (Hint: Compare $\sqrt{n}$ and $\sqrt{4n}$.)

2. A number is multiplied by 100. What happens to its square root?

3. A number is multiplied by 2. What happens to its square root?

4. A number is multiplied by 10. What happens to its square root?

5. The formula $r = \sqrt{12L}$ estimates the car speed r from length L of skid marks on wet pavement. Suppose the skid marks of one car are twice the length of another. What can you say about the speeds of the cars?

C 6. (A calculator with a square root key is helpful for this problem.) One of the fastest growing countries in the world is Brazil. In 1964, Brazil's population was estimated as 78,800,000. In 1972, an estimate was 99,000,000.

(a) Estimate the population in 1968 to the nearest 100,000.

(b) Use your answer to part (a) to estimate the 1970 population.

(c) Use your answer to part (b) to estimate the 1971 population.

(d) Use the 1971 and 1972 estimates to figure out the approximate rate of growth per year.

Lesson 6

Multiplying Binomials

A <u>binomial</u> is an expression of the form a + b. That is, a
binomial contains two terms which are added or subtracted. Here
are other binomials.

$$3x + \frac{y}{2} \qquad\qquad a - b \qquad\qquad 4 + 8m^3$$

The distributive property shows how to multiply a binomial by a
single number.

<u>Distributive Property</u>

$$(a + b)x = ax + bx$$

Length-area picture for the
<u>distributive property</u>

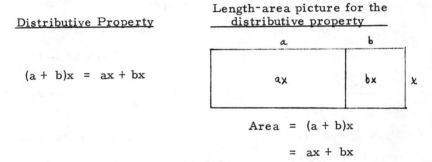

Area = (a + b)x

= ax + bx

Suppose x stands for a binomial. That is, substitute c + d for x.

<u>Product of two binomials</u>

$$(a + b)(c + d) = a(c + d) + b(c + d)$$

$$= ac + ad + bc + bd$$

Length-area picture for the
<u>product of two binomials</u>

Area = (a + b)(c + d)

= ac + ad + bc + bd

547

The result of multiplying two binomials is important. So we label it as a theorem. The name of the theorem may help you remember it. The face helps, too!

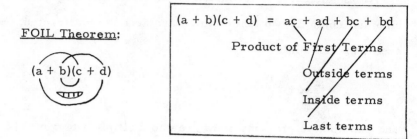

FOIL Theorem:

$(a + b)(c + d) = ac + ad + bc + bd$

Product of First Terms

Outside terms

Inside terms

Last terms

Examples: Applying the FOIL Theorem

1. $(x + 4)(x + 5)$

= $x^2 + 5x + 4x + 20$

= $x^2 + 9x + 20$

2. $(4t - 6)(2t + 3)$

= $8t^2 + 12t - 12t - 18$

= $8t^2 - 18$

3. $58 \cdot 75$

= $(50 + 8)(70 + 5)$

= $50 \cdot 70 + 50 \cdot 5 + 8 \cdot 70 + 8 \cdot 5$

= $3500 + 250 + 560 + 40$

= 4350

4. $(x^2 + 3)(a + b)$

= $x^2 a + x^2 b + 3a + 3b$

You can check your work by substitution. For instance, to check the answer to Example 1:

$$(x + 4)(x + 5) \stackrel{?}{=} x^2 + 9x + 20$$

Let x = 3: $(3 + 4)(3 + 5) \stackrel{?}{=} 3^2 + 9 \cdot 3 + 20$

548

$$7 \cdot 8 \overset{?}{=} 9 + 27 + 20$$

$$56 = 56 \qquad \text{Checks.}$$

Questions covering the reading

1. Give an example of a binomial.

2. Give an example of a trinomial.

3. What property of real numbers is most important for the FOIL Theorem?

4. Describe the FOIL Theorem. Explain how this theorem gets its name.

5. <u>Multiple Choice.</u> $(x - 5)(x - 3) =$

   (a) $x^2 - 15x - 8$ (b) $2x - 8$ (c) $x^2 - 8x + 15$ (d) none of these

6. How can you check an answer to a problem like that in Question 5?

7. Check that $(x + 3y)(2x - y) = 2x^2 + 5xy - 3y^2$.

8. What multiplication is pictured at right?

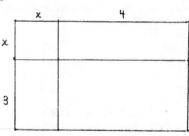

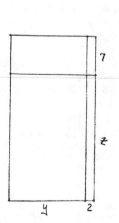

9. What multiplication is pictured at left?

549

10-25. Apply the FOIL Theorem.

10. $(x + 8)(x + 2)$

11. $(x + 1)(4 + x)$

12. $(y - 5)(y + 3)$

13. $(2 - m)(3 - m)$

14. $(2a + 4)(a - 3)$

15. $(3a + 6)(a - 2)$

16. $(b + 6)(b + 6)$

17. $(2a - 3b)(a + b)$

18. $(2x - 3)(2x + 3)$

19. $(m + 5)(22m + 5)$

20. $(x + y)(x + y)$

21. $(3 - a)(4 - 5a)$

22. $(6y - 2)(3y + 5)$

23. $(6B - 1)(2B + 4)$

24. $(^-x + 2)(^-x + 8)$

25. $(y + 1)(y + 2)$

26-31. Let $A = 3x - 2y$, $B = 2x + 3y$, $C = x - y$, and $D = 4x$. Calculate:

26. AB

27. BA

28. BD

29. BD + CD

30. $(B + C)D$

31. BCD

Questions extending the reading

1-4. An expression with three terms is called a <u>trinomial</u>. A trinomial has the form $a + b + c$. Notice how to multiply these expressions.

$$(a + b + c)(x + y) = a(x + y) + b(x + y) + c(x + y)$$

$$= ax + ay + bx + by + cx + cy$$

Multiply each term of the first factor by each term of the second factor. This generalizes the FOIL process. Apply this process to the problems below. In each case check your answer by letting the variable equal 2.

1. $(x^2 + x - 2)(x^2 - x - 2)$

2. $(3y^2 - 1)(2y^2 - 5)$

3. $(z - 1)(z^3 + z^2 + z + 1)$

4. $(4m^3 + 3m^2 + 2m + 1)(m + 1)$

5. Freshmen and sophomores are <u>underclassmen</u>; juniors and seniors are <u>upperclassmen</u>. Suppose one underclassman and one upperclassman are on a school discipline committee. How many such twosomes are possible in a school with:

> 550 freshmen
> 500 sophomores
> 450 juniors
> 400 seniors

6. Repeat Question 5 if there are f freshmen, p sophomores, j juniors, and s seniors.

7-8. True or False?

7. $56 \cdot 83 = 50 \cdot 80 + 6 \cdot 80 + 50 \cdot 3 + 6 \cdot 3$

8. $4\frac{1}{2} \cdot 9\frac{3}{4} = 4 \cdot 9 + 4 \cdot \frac{3}{4} + \frac{1}{2} \cdot 9 + \frac{1}{2} \cdot \frac{3}{4}$

9-10. Use the idea of Question 8 to multiply.

9. $12\frac{1}{3} \cdot 6\frac{1}{3}$

10. $7\frac{2}{7} \cdot 6\frac{5}{7}$

11. (a) What multiplication problem can be pictured at right?

    (b) What theorem is shown by the picture?

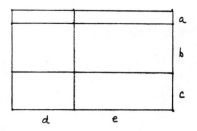

12. (a) What multiplication problem can be pictured at left?

    (b) What theorem is shown by the picture?

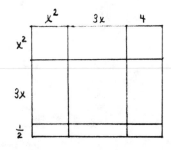

13. Draw a length-area picture for the theorem

$$(a + b + c)(d + e + f) = ad + ae + af + bd + be + bf + cd + ce + cf.$$

14. Why are both length and area needed in the picture for the multi-plication of two binomials?

Arithmetic Shortcuts

At left some integers are multiplied. At right the same integers are written in a way which shows the pattern of the answers.

$$50 \cdot 50 = 2500 \qquad (50 + 0)(50 - 0) = 2500 - 0$$
$$51 \cdot 49 = 2499 \qquad (50 + 1)(50 - 1) = 2500 - 1$$
$$52 \cdot 48 = 2496 \qquad (50 + 2)(50 - 2) = 2500 - 4$$
$$53 \cdot 47 = 2491 \qquad (50 + 3)(50 - 3) = 2500 - 9$$
$$54 \cdot 46 = 2484 \qquad (50 + 4)(50 - 4) = 2500 - 16$$
$$\vdots \qquad\qquad\qquad\qquad \vdots$$

The numbers on the far right are all squares. This gives a hint to the pattern.

$$(50 + n)(50 - n) = 50^2 - n^2$$

You can check this pattern by using the FOIL Theorem.

$$(50 + n)(50 - n) = 50 \cdot 50 - 50n + 50n - n^2$$
$$= 50^2 - n^2$$

Actually a more general pattern is true:

$$(x + y)(x - y) = x^2 - y^2$$

The pattern is so simple that you could do all of the above multiplications in your head. Here is the shortcut.

Arithmetic Shortcut 1: To multiply two integers:

(1) Find their mean. If the integers are 53 and 47, the mean is 50.

(2) Find the distance from either number to the mean. Distance from 53 to 50 is 3.

(3) Subtract the squares of the mean and the distance. $53 \cdot 47 = 50^2 - 3^2 = 2491$

This shortcut works best when you can find the mean with no work and when the mean ends in 0. For example, to multiply 41 by 39: (1) The mean is 40. (2) The distance is 1. (3) The product is $40^2 - 1^2$ or 1599. What you have really done is use the idea that $(40 + 1)(40 - 1) = 40^2 - 1^2$.

A second shortcut involves the square of a sum. In general,

$$(x + y)^2 = (x + y)(x + y)$$
$$= x^2 + xy + yx + y^2$$
$$= x^2 + 2xy + y^2$$

When $y = 1$: $\qquad (x + 1)^2 = x^2 + 2x + 1$

If x is an integer, $x + 1$ is the next integer after x. This leads to another shortcut.

Arithmetic Shortcut 2: Suppose you know the square of an integer x.
Add $2x + 1$ to that square.
The result is the square of $x + 1$.

For example, you know $100^2 = 10000$. Let x be 100. Then $2x + 1 = 201$.

$$(x + 1)^2 = x^2 + 2x + 1$$
$$101^2 = 100^2 + 2 \cdot 100 + 1$$
$$= 10000 + 201$$
$$= 10201$$

This could be done in your head. That's its advantage over multiplying 101 by 101 using paper and pencil or a calculator.

You can also use this idea to calculate numbers like $(1.08)^2$ in your head. Think of 1.08 as $1 + .08$.

$$(x + y)^2 = x^2 + 2xy + y^2$$
$$(1 + .08)^2 = 1^2 + 2 \cdot 1 \cdot .08 + (.08)^2$$
$$= 1 + .16 + .0064$$
$$= 1.1664$$

The products $(x + y)(x - y)$ and $(x + y)(x + y)$ lead to shortcuts. The product $(x - y)(x - y)$ also can lead to shortcuts (see Question 2, page 557) and is very important in applications.

$$(x - y)(x - y) = x^2 - xy - yx + y^2$$
$$= x^2 - 2xy + y^2$$

These three products are so important that you might want to memorize them. But if you do not memorize them, you can always use the FOIL Theorem to get them from scratch.

Special Products Theorem:		
	Square of Sum:	$(x + y)^2 = x^2 + 2xy + y^2$
	Square of Difference:	$(x - y)^2 = x^2 - 2xy + y^2$
	Difference of Squares:	$(x + y)(x - y) = x^2 - y^2$

Examples: Special products

Square of sum: 1. $(x + 3)^2 = x^2 + 6x + 9$

2. $(10t + u)^2 = 100t^2 + 20tu + u^2$

Square of difference:

3. $(a - 30)^2 = a^2 - 60a + 900$

4. $(7B - 6)^2 = 49B^2 - 84B + 36$

Difference of squares:

5. $(w + 5)(w - 5) = w^2 - 25$

6. $(3a - 1)(3a + 1) = 9a^2 - 1$

Questions covering the reading

1-15.   Give the answers to these multiplication problems using Shortcut 1 in this lesson.

1. $60 \cdot 60$ 2. $61 \cdot 59$ 3. $62 \cdot 58$ 4. $63 \cdot 57$ 5. $64 \cdot 56$

6. $70 \cdot 70$ 7. $73 \cdot 67$ 8. $75 \cdot 65$ 9. $20 \cdot 20$ 10. $19 \cdot 21$

11. $17 \cdot 23$ 12. $90 \cdot 90$ 13. $85 \cdot 95$ 14. $88 \cdot 92$ 15. $39 \cdot 41$

16-35.   Write as a polynomial.

16. $(a - b)(a + b)$ 17. $(3 + x)^2$ 18. $(3 - x)^2$

19. $(x - 6)(x + 6)$ 20. $(a + 4)^2$ 21. $(z - 5)^2$

22. $(2y + 5)(2y - 5)$ 23. $(3m - 2)^2$ 24. $(n + 2)^2$

25. $(1 - y)(1 + y)$ 26. $(\frac{z + 3}{2})^2$ 27. $(3p - 1)^2$

28. $(\frac{1}{2} + 3m^2)(\frac{1}{2} - 3m^2)$ 29. $3(2x - 3)^2$ 30. $4(9 + y)^2$

31. $(a + 7)^2 - (a - 7)^2$ 32. $10(B - 6)^2$ 33. $\frac{1}{2}(a + 2)^2$

34. $x^2 + y^2 - (x + y)^2$ 35. $(2m + n)(2m - n) + (2n - m)(2n + m)$

36. How much larger is $(x + 1)^2$ than x^2?

37. $40^2 = 1600$ so $41^2 = $ _____.

38. $25^2 = 625$ so $26^2 = $ _____.

39. $90^2 = 8100$ so $91^2 = $ _____.

40. $60^2 = 3600$ so $61^2 = $ _____.

41-44. Square each number without multiplying.

41. 1.02   42. 1.03   43. 1.09   44. 1.05

Questions for exploration

1. Look in the table of squares, pages 742 and 743, and write the squares of all numbers which end in 5. Try to find a pattern which would enable you to calculate the squares in your head.

2. Notice that $(x - 1)^2 = x^2 - 2x + 1$. This pattern is similar to arithmetic shortcut 2. It can be used to square numbers mentally which end in 9. (a) What is the shortcut? (b) Apply the shortcut to calculate 79^2 (knowing that $80^2 = 6400$).

3. Write $(x + y)^3$ as a polynomial. (Remember:ʳ $(x + y)^3 \neq x^3 + y^3$.)

4. Write $(x - y)^3$ as a polynomial.

Lesson 8

Geometric Shortcuts and the Pythagorean Theorem

Roads $\overline{AC}$ and $\overline{BC}$ meet at right angles.

Some hikers wish to go from point A to point B. Instead of walking 5 km along the roads, they decide to cross the field. How long is the shortcut $\overline{AB}$?

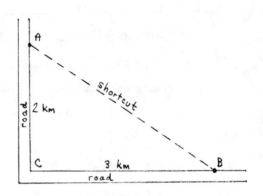

You can find the length of $\overline{AB}$ by a process first used in Lesson 1.

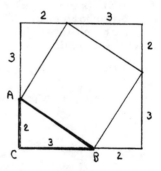

1. Draw a right triangle with legs 2 and 3.

2. Place 3 triangles with identical dimensions as at right.

3. The symmetric arrangement of the triangles forms a tilted square in the middle.

Area of tilted square = Area of large square - 4(Area of ABC)

$$AB^2 = \qquad 25 \qquad - 4 \cdot 3$$

$$AB^2 = 13$$

4. By the definition of square root,

$$AB = \sqrt{13}$$

The shortcut has length $\sqrt{13}$ or about 3.6 km. This saves about 1.4 km of walking.

You could go through this process every time you wanted to find the length of a shortcut. But it is easier to use a general formula. This formula can be found by going through the above process with variables in place of the numbers.

Given: Segments with lengths x and y which form a right angle.

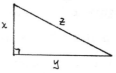

We want to find how x, y, and z are related.

Step 1: Draw triangles, form-ing the tilted square just as before.

Step 2: The large square has side x + y and so has area $(x + y)^2$.

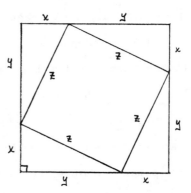

Step 3:

Area of tilted square = Area of large square - 4(Area of triangle)

Substituting: $z^2 = (x + y)^2 - 4(\frac{1}{2}xy)$

Step 4: $z^2 = x^2 + 2xy + y^2 - 2xy$

Simplifying: $z^2 = x^2 + y^2$

Step 5: $z = \sqrt{x^2 + y^2}$

This result is known as the Pythagorean Theorem.

<u>Pythagorean Theorem:</u>

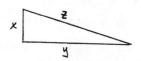

> In a right triangle with sides x, y, and z, z being the longest side,
>
> $$z^2 = x^2 + y^2$$
>
> That is, $\qquad z = \sqrt{x^2 + y^2}$

The Pythagorean Theorem is probably <u>the most famous theorem</u> in all of mathematics. It is named after the group of mathematicians and mystics that was founded by Pythagoras in the 6th century B.C.

Examples: Applying the Pythagorean Theorem

1. In the hiking problem x = 2 and y = 3. By the Pythagorean Theorem, the shortcut has length $\sqrt{2^2 + 3^2}$ or $\sqrt{4 + 9}$ or $\sqrt{13}$. This checks with what was found on the previous page.

2. The longest segment that can be drawn on a sheet of $8\frac{1}{2}$ by 11" note- book paper is pictured by the diagonal drawn in at right. It's

    length is $\sqrt{(8\frac{1}{2})^2 + 11^2}$ = $\sqrt{72.25 + 121}$ = $\sqrt{193.25}$ ≈ 13.9"

3. By the Pythagorean Theorem,

    AC = $\sqrt{6^2 + 8^2}$

    = $\sqrt{36 + 64}$

    = $\sqrt{100}$ = 10

Pythagoras and his followers lived in what is now Italy. Then much of Italy was part of the Greek empire, so Pythagoras is called Greek. There are examples of the Pythagorean Theorem in very old Egyptian and Babylonian writings. But the Pythagoreans probably were the first to prove the general result. The method they used to prove the theorem was geometric. They could not prove the theorem as we did because algebra was not known to the Greeks. The use of algebra to gain information about geometric ideas and pictures was not known until the 1600's, over 2000 years after Pythagoras.

Questions covering the reading

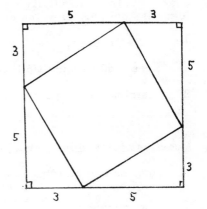

1. Use the drawing at right.

   (a) What is the area of the big square?

   (b) What is the area of each triangle?

   (c) What is the area of the tilted square?

   (d) How long is a side of the tilted square?

2. Repeat Question 1 if the sides of the triangle are a and b.

3. State the Pythagorean Theorem.

4. If Pythagoras lived in what is now Italy, why is he called a Greek geometer?

5. When did Pythagoras live?

6. Did Pythagoras know algebra?

7-15. The two shorter sides of a right triangle have the given lengths. (a) Find the length of the longest side to the nearest tenth. (b) At least once, check your answer by making a careful drawing.

7. 3 and 4

8. 7 and 24

9. 3 and 3

10. 5 and 5

11. 1 and 2

12. .8 and .6

13. 2 and 2.1

14. 3/4 and 11/4

15. 8/7 and 15/7

16. What is the length of the longest segment that can be drawn on a 3" by 5" index card?

17. How long is a diagonal of a rectangle which is 7.5 cm long and 10 cm wide?

18. How long is a diagonal of a square with side z?

19-23. In the general argument on page 559, each of the following ideas is applied. In which step is each idea used?

19. FOIL Theorem

20. Cutting off model of subtraction

21. Area model of multiplication

22. Definition of square root

23. Distributive property

Questions testing understanding of the reading

1. Suppose one building is 2 km east and 4 km south of a second building. What is the helicopter distance between the buildings?

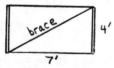

2. How long must the brace be for a fence gate 4' high and 7' wide?

3. How much distance is saved by going diagonally across a square km of farmland instead of going along the sides?

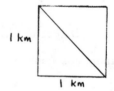

4. The dimensions of flat rectangular boxes are given. In which box can a meter stick be shipped?

    (a) 50 cm by 80 cm   (b) 65 cm by 65 cm

    (c) 90 cm by 40 cm   (d) none of these

5. If $x^2 + y^2 = z^2$ and z is positive, which is <u>not</u> true?

    (a) $x + y = z$ (b) $\sqrt{x^2 + y^2} = z$ (c) $x^2 = z^2 - y^2$

6-11. Use graph paper if necessary. Find the distance between the points. (Hint: Draw a right triangle on the graph paper.)

6. (0, 0) and (-3, 4)   7. (1, 2) and (2, 3)

8. (1, 5) and (5, 2)   9. (0, 0) and (-70, -20)

10. (-3, 5) and (6, -4)   11. (1.3, 2.1) and (4.8, -9.9)

Lesson 9 (Optional)

A Statistic - The Standard Deviation

Review: In Chapter 3 you learned about a statistic called the mean absolute deviation, or m. a. d. The m. a. d. measures dispersion, how spread out some given numbers are. The more spread out or dispersed the numbers, the higher the m. a. d.

A more common statistic for measuring dispersion is called the standard deviation, abbreviated s. d. Here is how to calculate the s. d.

Algorithm for calculating standard deviation	Example of calculation of standard deviation
Step 1: Find the mean of the numbers. Call it m.	1. If the numbers are 30, 40, 50, and 60, m = 45.
Step 2: Subtract m from each number. (The differences you get are the deviations.)	2. The differences are $$30 - 45 = -15$$ $$40 - 45 = -5$$ $$50 - 45 = 5$$ $$60 - 45 = 15$$
Step 3: Square the differences.	3. The squared differences are 225, 25, 25, and 225.
Step 4: Find the mean of the squared differences.	4. The mean is $$\frac{225 + 25 + 25 + 225}{4} = 125$$
Step 5: Take the square root of the mean. This is the standard deviation.	5. s. d. $= \sqrt{125}$ $$\approx 11.2$$

It may be easier for you to remember a formula for s.d. rather than the 5-step algorithm.

Definition of
Standard
Deviation:

> Let there be n given numbers called S_1, S_2,
> ..., S_n whose mean is m. Then the standard
> deviation (s.d.) of these numbers is:
>
> $$s.d. = \sqrt{\frac{(S_1 - m)^2 + (S_2 - m)^2 + \ldots + (S_n - m)^2}{n}}$$

Second example: Calculate the s.d. of 35, 60, 45, 50, and 60.

Calculation: S_1 = 35, S_2 = 60, S_3 = 45, S_4 = 50, S_5 = 60.

The number of numbers is 5.   n = 5.

The mean is 50.   m = 50

$$s.d. = \sqrt{\frac{(35-50)^2 + (60-50)^2 + (45-50)^2 + (50-50)^2 + (60-50)^2}{5}}$$

$$= \sqrt{\frac{225 + 100 + 25 + 0 + 100}{5}} = \sqrt{\frac{450}{5}} = \sqrt{90} \approx 9.49$$

In the next example, there are again 5 numbers. The mean is again 50. But these numbers are more widely dispersed than those in the second example. So you should expect a larger s.d.

Third example: Calculate the s.d. of 10, 100, 60, 35, and 45.

Calculation: S_1 = 10, S_2 = 100, S_3 = 60, S_4 = 35, and S_5 = 45.

n = 5 and m = 50.

$$s.d. = \sqrt{\frac{(10-50)^2 + (100-50)^2 + (60-50)^2 + (35-50)^2 + (45-50)^2}{5}}$$

$$= \sqrt{\frac{1600 + 2500 + 100 + 225 + 25}{5}} = \sqrt{\frac{4450}{5}} = \sqrt{890} \approx 29.8$$

The name "standard deviation" was first used in 1894 by Karl Pearson. Pearson, a British mathematician, was born in 1857 and died in 1936. He was one of the founders of modern statistics.

Questions covering the reading

1-10. Calculate the standard deviation of each collection of scores.

1. 10, 20, 30, 40

2. 70, 68, 62, 80, 75

3. 27, 25, 23, 25

4. 170, 168, 162, 180, 175

5. 3, 3, 4, 6, 5, 3

6. 10, 8, 8, 8, 10, 16

7. 90, 110, 90, 110, 90, 110, 90, 110

8. 96, 96, 96, 96, 96, 96, 96, 96

C 9. 40, 44, 50, 44, 50, 56, 56, 60

C 10. 30, 38, 50, 38, 50, 62, 62, 70

11. Why should you expect the answer to Question 3 to be less than the answer to Question 1?

12. Why should you expect the answers to Questions 2 and 4 to be equal?

13. In the formula for calculating the s.d.:

(a) What is n?   (b) What are S_1, S_2, ..., S_n?

(c) What is m?

14. Who first used the term "standard deviation?"

15. Which statistic is more commonly used, the m. a. d. or the s. d. ?

16-19. Review.

16. What is an algorithm?

17. What does the word "dispersion" mean?

18. Calculate the m. a. d. for the scores of Question 2.

19. Calculate the m. a. d. for the scores of Question 1.

20. Part of the algorithm for calculating the s. d. is called the "root mean square." Which steps does this part include?

Questions applying the reading

1-4. Which class, A or B, was more consistent on the test?

1.   Class A: Mean 86; s. d. 10. 3
     Class B: Mean 84; s. d. 11. 0

2.   Class A: Mean 72; s. d. 6. 3
     Class B: Mean 90; s. d. 6. 3

3.   Class A: Mean 11. 2; s. d. 2. 34
     Class B: Mean 12. 1; s. d. 2. 38

4.   Class A: Mean 50; s. d. 10
     Class B: Mean 50; s. d. 8

5-6.   Data given is average monthly temperature for years 1931-1970. Source: National Climatic Center. Use the s. d. to help answer the given question.

C 5.   Which city has the more consistent temperature?

	Jan	Feb	M	A	M	J	J	A	S	O	N	D
Albany:	22	24	33	46	58	67	72	70	62	51	39	26
Chicago:	26	28	36	49	60	70	76	74	66	55	40	29

C 6.   Which city has the more consistent temperature?

| Houston: | 54 | 56 | 61 | 68 | 76 | 82 | 83 | 83 | 79 | 71 | 61 | 56 |
| Mobile: | 53 | 55 | 60 | 68 | 76 | 82 | 82 | 82 | 78 | 70 | 59 | 54 |

There are three ways to think about square roots. First, if $x^2 = A$, then x is called a square root of A. The square roots of A are written $\sqrt{A}$ and $-\sqrt{A}$.

Second, if a square has area A, then its sides have length $\sqrt{A}$. The area view of square roots is connected to the area model of multiplication. That model leads to two very important theorems:

FOIL Theorem: $(a + b)(c + d) = ac + ad + bc + bd$

Pythagorean Theorem: In a right triangle with shorter sides of lengths x and y, the longest side z has length $\sqrt{x^2 + y^2}$, and $z^2 = x^2 + y^2$

The FOIL Theorem enables any polynomials to be multiplied and leads to some nice arithmetic shortcuts. The Pythagorean Theorem has many applications to geometrical problems.

A third view of square roots is related to the growth model of powering. If a quantity is multiplied by B in a time period, then in half that time, it will be multiplied by $B^{1/2}$. It is easy to see that

$B^{1/2} \cdot B^{1/2} = B$, so $B^{1/2}$ must be equal to $\sqrt{B}$.

In general, if a quantity grows exponentially--that is, by a scale factor of some fixed amount during time periods of the same length-- and there is x at the beginning of the period and y at the end, then there will be $\sqrt{xy}$ in the middle. $\sqrt{xy}$ is called the geometric mean of x and y and is equal to $\sqrt{x} \cdot \sqrt{y}$. This property of the geometric mean, a special case of $(xy)^n = x^n y^n$, helps to simplify many expressions that have square roots.

A further application of square roots is in the calculation of standard deviation, a statistic commonly used to measure dispersion.

CHAPTER 13

SETS AND EVENTS

Lesson 1

Sets, Subsets, and Elements

A <u>set</u> is a collection of objects, called <u>elements</u>. Usually the elements are put together for a purpose. Sets are found outside of mathematics, but usually with a different name.

<u>When a set is called</u>	<u>Element is often called</u>
team	player
committee	member
the U.S. Senate	senator
class	student
herd of cattle	cow
place setting	dish

A set often has different properties than its elements. For example:

1. A team in baseball can win the World Series, a player cannot.
2. A player can hit a home run, a team cannot.
3. The Senate can pass legislation, a senator cannot.
4. A cow can be milked, a herd cannot.

The standard symbol used for a set is $\{\ldots\}$, and commas go

Nell,
between the elements. When Lulu, Mike, Oscar, Paula, and Quincy
are the 6 members of a committee, you could write:

committee = $\Big\{$ Lulu, Mike, Nell, Oscar, Paula, Quincy $\Big\}$

The order of naming elements makes no difference. Here is the
same committee.

= $\Big\{$ Oscar, Mike, Lulu, Nell, Paula, Quincy $\Big\}$

Two sets are <u>equal</u> if they have the same elements.

A committee may have subcommittees. So a set has <u>subsets</u>.
One possible subcommittee is

$\Big\{$ Oscar, Quincy, Paula $\Big\}$

In short, $\{O, Q, P\}$ is a subset of $\{L, M, N, O, P, Q\}$. In
general, set ① is a subset of set ② if every element of set ①
is also an element of set ② .

Examples: Sets, subsets, and elements

<u>Set</u>	Example of an <u>element of the set</u>	Example of a <u>subset of the set</u>
1. set of integers	eight	set of even integers
$\{0, 1, -1, 2, -2, \ldots\}$	8	$\{0, 2, -2, 4, -4, \ldots\}$
2. $\{x, x^2, x^3, \ldots\}$	x^3	$\{x^{12}, x^{10}\}$
3. triangle ABC	point A	segment $\overline{AC}$

571

There is a set which has no elements in it. It is called the empty set or null set. The symbol $\{\ \}$ is naturally used to refer to this set. The Danish letter $\emptyset$ is also used. The set $\emptyset$ occurs often in situations. It might refer to

(1) the set of points of intersection of 2 parallel lines,

(2) the set of boys on a girls basketball team,

etc.

Sets were first used in mathematics in the 1800's. In the late 1950's and 1960's, sets became identified with "new math." Actually, much mathematics is newer than sets. For example, most of statistics and almost all of computer mathematics has been discovered and developed in this century.

Questions covering the reading

1. The objects in a set are called _____.

2-8. A set and an element are given. (a) Name another element of the set. (b) Give an example of a subset of the set.

2. set: baseball team element: pitcher

3. set: committee element: committee chairperson

4. set: family element: mother

5. set: $\{2,\ 11,\ -6\}$ element: -6

6. set of real numbers element: 2

7. $\{x,\ x^2,\ x^3,\ \ldots\}$ element: x^{10}

8. set: square ABCD   element: point M

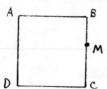

9-14. The given set is a subset of one of the sets given in Questions
2-8. Name the larger set.

9. set of parents   10. $\{2,\ 5,\ -4.\ 3\}$

11. set of infielders   12. subcommittee

13. $\overline{AB}$ 14. $\{5\}$

15-18. Identify the set, the subset, and the element.

15. Senate, Foreign Relations committee, senator

16. place setting, service for eight, cup

17. student, row nearest the window, class

18. Michiganders, U.S. citizens, Henry Ford

19. Give an example of (a) a set S and (b) a subset of S which has
     exactly 3 elements.

20. Give an example of (a) a set S and (b) a subset of S which has
     exactly 100 elements.

21. What does $\emptyset$ stand for?   22. What does $\{\ \}$ stand for?

23. Give a situation in which $\emptyset$ might be used.

24. Why does $\{\emptyset\}$ not stand for the empty set?

25-28. A set and an element are given. (a) Name something which
can apply to the set but not its elements. (b) Name something which
can apply to the elements but not to the set.

25. set: committee   element: member

26. set: senate   element: senator

27. set: football team   element: quarterback

28. set: herd   element: cow

Questions extending the reading

1-4. A set is called <u>infinite</u> if it has infinitely many elements. Otherwise, it is called <u>finite</u>. Refer to Questions 2-14 on page 572.

1. Which sets of Questions 2-8 are infinite?

2. Which sets of Questions 9-14 are infinite?

3. Which sets of Questions 2-8 are finite?

4. Which sets of Questions 9-14 are finite?

5-6. A set is called a <u>singleton</u> if it has exactly one element.

5. Give an example of a set which is a singleton.

6. Which sets of Questions 2-14, page 572, are singletons?

7-12. A set of numbers is <u>closed under addition</u> if when you add <u>any</u> two elements of the set, the sum is in the set.

Examples: the set of positive integers is closed under addition.

$$\left\{\frac{1}{2},\ 1,\ 2,\ 3,\ 4,\ \ldots\right\} \text{ is not closed under addition}$$

$$\text{because } 1 + \frac{1}{2} \text{ is not in the set.}$$

Is the given set closed under addition?

7. the set of negative real numbers   8. $\{2, 4, 6, 8, 10, \ldots\}$

9. the set of positive real numbers   10. $\{1, 3, 5, 7, 9, \ldots\}$

11. $\{0\}$ 12. $\{0, 1\}$

13. Make up a definition of <u>closed under multiplication</u>. (See Questions 7-12 for a hint.)

14. Which of the sets in Questions 7-12 above are closed under multiplication?

<u>Question for exploration</u>

1. A set of cattle is called a <u>herd</u>. Give a name which can refer to
    a set of:

    (a) sheep   (b) bees   (c) lions

    (d) wolves   (e) elephants   (f) fish

    (g) geese   (h) quail

The Abbreviations P(E) and N(E)

In Lesson 3 of Chapter 5, starting on page 226, three important statements were made.

1. An event is a set of outcomes.

2. The probability of an event is the sum of the probabilities of its outcomes.

3. If the outcomes in an experiment occur randomly, then the probability of an event

$$= \frac{\text{number of outcomes in the event}}{\text{number of outcomes in the experiment}}$$

If you have forgotten what these statements mean, you might want to reread that lesson. Here is an example not given in that lesson.

Experiment: A fair coin is tossed 3 times.

Question: What is the probability of getting exactly 2 heads in the 3 tosses?

Solution: This experiment has 8 outcomes. They are listed here.

HHH	THH
HHT	THT
HTH	TTH
HTT	TTT

The event getting two heads =

$\{$HHT, HTH, THH$\}$. Since that

event has 3 outcomes and there

are 8 outcomes in all, the proba-

bility of the event is $\frac{3}{8}$.

In calculating probabilities, two long English phrases occur

often. So it is helpful to have abbreviations.

English phrase	mathematical abbreviation
the probability of the event E	P(E)
the number of elements in the set E	N(E)

In the coin example, if   E = "getting two heads," then

$$P(E) \ = \ \frac{3}{8}$$

$$N(E) \ = \ 3$$

One other definition is very helpful. The set of all outcomes to

an experiment is called the sample space of the experiment. For the

coin experiment, the sample space contained 8 outcomes. These

outcomes are elements of the sample space. If the sample space is

named S, then

$$N(S) \ = \ 8$$

With these abbreviations, Statement 3 on page 576 (the Proba-

bility of an Event Theorem) can now be written:

> Let E be an event in an experiment with
> sample space S and assume outcomes in S
> occur randomly. Then
> $$P(E) = \frac{N(E)}{N(S)}.$$

Even though there are parentheses in P(E) and N(E), neither P(E) nor N(E) has anything to do with multiplication. They are just abbreviations.

The next example seems different from coin-tossing but mathematically uses the same idea.

Experiment: You randomly guess at five true-false questions on a test.

Question: What is the probability that you will pass with a grade of 70% or higher?

Solution: For each question, you can be right (R) or wrong (W). Since there are 5 questions, there are 2^5 or 32 outcomes in the sample space. They are listed here.

RRRRR	RWRRR	WRRRR	WWRRR
RRRRW	RWRRW	WRRRW	WWRRW
RRRWR	RWRWR	WRRWR	WWRWR
RRRWW	RWRWW	WRRWW	WWRWW
RRWRR	RWWRR	WRWRR	WWWRR
RRWRW	RWWRW	WRWRW	WWWRW
RRWWR	RWWWR	WRWWR	WWWWR
RRWWW	RWWWW	WRWWW	WWWWW

3 right is 60%, 4 right is 80%, so you need
4 or 5 right to pass.

So passing = {RRRRR, RRRRW, RRRWR, RRWRR,

RWRRR, WRRRR}

$$P(\text{passing}) = \frac{N(\text{passing})}{N(\text{Sample Space})} = \frac{6}{32}$$

$\frac{6}{32}$ = .1875; you would have less than a 19%
chance of passing.

Questions covering the reading

1-6. Give the definition for each term or symbol.

1. event   2. sample space

3. N(E)   4. P(E)

5. probability of an event   6. Review. random outcomes

7-13. A fair coin is tossed 3 times.

7. Write down all elements in the sample space.

8. Write down all elements in the event "getting 2 heads."

9. Calculate N(getting 2 heads).

10. Calculate P(getting 2 heads).

11. Name all elements in the event "getting 3 heads."

12. Calculate N(getting 3 heads). 13. Calculate P(getting 3 heads).

14. State the Probability of an Event Theorem.

15-23. A true-false test has 5 questions. A person randomly guesses on all of them.

Let S = sample space E_1 = "getting 1 question correct"

　　Z = "passing with a E_2 = "getting 2 questions correct"
　　　　grade over 70%"

　　　　　　　　　　　　　　E_3 = "getting 3 questions correct"

　　　　　　　　　　　　　　E_4 = "getting 4 questions correct"

　　　　　　　　　　　　　　E_5 = "getting 5 questions correct"

Calculate:

15. N(S)	16. N(Z)	17. P(Z)
18. $N(E_1)$	19. $P(E_1)$	20. $P(E_2)$
21. $N(E_2)$	22. $P(E_3)$	23. $P(E_4)$

Questions testing understanding of the reading

1. Match the terms at left with those at right.

　　(a) event   (x) set
　　(b) outcome   (y) subset
　　(c) sample space   (z) element

2. Multiple Choice. Refer to the coin example on page 576. If the probability of heads in a single toss is not 1/2, which of these would change and why?

　　(a) N (getting two heads)   (b) N (sample space)
　　(c) P (getting two heads)

580

3-6. The sex of a newborn child can almost be considered to occur randomly. Assuming this, suppose a family has 3 children.

3. What is the probability that exactly 2 are boys?

4. What is the probability that exactly 2 are girls?

5. What is the probability that all are girls?

6. What is the probability that all are of the same sex?

7. A fair coin is tossed twice. (a) Name all elements of the sample space. (b) What is the probability of getting exactly 1 head?

8. A fair coin is tossed 4 times. (a) Write down all elements of the sample space. (b) What is the probability of getting exactly 2 heads? (c) What is the probability of getting 4 tails?

9. A fair coin is tossed 10 times. (a) How many elements are in the sample space? (b) What is the probability of all heads?

10. A fair coin is tossed n times. (a) How many elements are in the sample space? (b) What is the probability of all heads?

11-14. A fair coin is tossed 5 times. Choose the more common event.

11. Getting exactly 1 head or getting exactly 2 heads.

12. Getting exactly 2 heads or getting exactly 3 heads.

13. Getting exactly 3 heads or getting exactly 4 heads.

14. Getting exactly 4 heads or getting exactly 5 heads.

Lesson 3

Expected Frequencies

Imagine tossing a fair coin 500 times. When the coin is fair, you expect heads half of the time. You would expect 250 heads. Of course, you wouldn't always get exactly 250 heads even if the coin were fair. Getting 249 or 251 or 240 or 260 would be almost as likely. But we call 250 the expected frequency of heads.

According to the Census Bureau 1973 data, about 5.0% of U.S. families have 7 or more persons in them. If you picked 30 families at random, you would expect 5% of them to have 7 or more persons. The expected frequency is 5% • 30 or 1.50. In this case, the expected frequency is not an integer. In other words, you could not get the expected frequency.

These examples suggest the following general definition.

Definition

> Suppose an event has probability p.
> Then if an experiment is repeated
> n times, the expected frequency of
> the event is np.

Here is a third example. The experiment is the tossing of 2 coins and we are interested in the number of heads which appear. There are four possible outcomes.

HH   HT   TH   TT

If the coin is fair, each outcome has probability 1/4. Then

$$P(0 \text{ heads}) = P(\{TT\}) = 1/4$$

$$P(1 \text{ head}) = P(\{HT, TH\}) = 1/2$$

$$P(2 \text{ heads}) = P(\{HH\}) = 1/4$$

Now think of repeating this experiment 100 times.

Expected frequency of 0 heads = $100 \cdot 1/4 = 25$

Expected frequency of 1 head = $100 \cdot 1/2 = 50$

Expected frequency of 2 heads = $100 \cdot 1/4 = 25$

As a final example, think of throwing two six-sided dice. Usually you add the numbers which appear on the dice. Let $P(t)$ = the probability that the sum is t.

For two dice, it is easy to calculate $P(t)$. This was done on page 228. The answers are at left. Corresponding points are graphed at right.

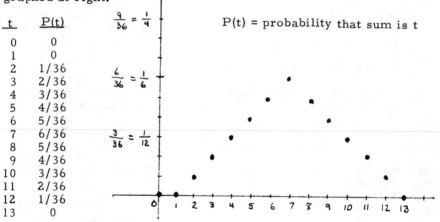

t	P(t)
0	0
1	0
2	1/36
3	2/36
4	3/36
5	4/36
6	5/36
7	6/36
8	5/36
9	4/36
10	3/36
11	2/36
12	1/36
13	0

$P(t)$ = probability that sum is t

Since the values of $P(t)$
add to 1, there are no more
possibilities.

583

The probabilities on page 583 help answer questions about expected frequencies.

Question: If two fair dice are thrown 50 times, how many 7's are expected?

Answer: Expected frequency of 7's = 50 · P(7)

$$= 50 \cdot \frac{6}{36} = 50 \cdot \frac{1}{6} = 8.1\overline{6}$$

You should expect about 8 7's.

Questions covering the reading

1. Imagine tossing a fair coin 30 times. How many tails would you expect? How many heads would you expect?

2. Suppose that 2/3 of the students in a school take a bus to school. In a class of 50 students, how many would you expect to have taken a bus? How many would you expect to have used some other way to get to school?

3. Suppose you toss a fair coin n times. How many heads would you expect?

4. In a school with s students, if the probability that a student will take the bus is .63, how many students should be expected to take the bus?

5. What is the expected frequency of an event?

6. Suppose F families are picked at random from the United States. What is the expected frequency of families with 7 or more persons?

7-9. Two fair coins are tossed 20 times.

7. (a) What is the probability of getting 2 heads on a particular toss?
   (b) What is the expected frequency of 2 heads?

8. (a) What is the probability of getting 1 head and 1 tail on a particular toss?

(b) What is the expected frequency of 1 head and 1 tail?

9. (a) What is the probability of getting 2 tails on a particular toss?
   (b) What is the expected frequency of 2 tails?

10-17. Two fair dice are tossed. For each event, give (a) the
probability and (b) the expected frequency in 100 tosses of 2 dice.

10. a sum of 2   11. a sum of 12

12. a sum of 3   13. a sum of 11

14. a sum of 6   15. a sum of 8

16. a sum of 7   17. a sum of 17

18. Suppose that you think you have a fair coin. You toss it 500
    times and heads occurs 245 times. Which is true?

    (a) The coin is definitely not a fair coin.

    (b) The coin is a fair coin.

    (c) You can't tell whether the coin is fair or unfair from this
        information but the evidence is in favor of it being fair.

    (d) You can't tell whether the coin is fair or unfair from this
        information but the evidence is against it being fair.

19. Repeat Question 18 if you get 200 heads in 500 tosses.

Questions testing understanding of the reading

1. Suppose two evenly matched teams meet in a "double-header."
   (They play two games against each other.)

   (a) What is the probability that each team will win one game?

   (b) Over a season with 5 double-headers, what is the expected
       frequency of double headers in which each team wins one
       game?

585

2-5. The four major blood types are A, B, AB, and O. A person can give blood to another person of the same type but not always to people of other types. The diagram summarizes the possibilities.

The arrow means
"can give to."

$$O \to A, \quad O \to B, \quad A \to AB, \quad B \to AB$$

(There are other factors which affect the ability to give or receive blood. These factors are ignored in the questions which follow.)

By sampling from the three populations named below, probabilities have been determined for the four major blood types. Probabilities for a fourth unknown population are also given. It is helpful for hospitals to know these probabilities so that they can determine whether there is enough blood on hand or available.

Blood Type	U.S. Black	U.S. White	West African	Unknown
O	.47	.45	.46	p
A	.28	.40	.23	q
B	.20	.11	.26	r
AB	.05	.04	.05	s

2. In a population of 100,000 U.S. Blacks, how many would be expected to:

   (a) be of blood type A.

   (b) be able to receive type A blood.

   (c) be able to give blood to a person of blood type A.

3. Repeat Question 2 for a population of 100,000 U.S. Whites.

4. Repeat Question 2 for a population of 100,000 West Africans.

5. Repeat Question 2 for 100,000 people from a place where the distribution of blood types is unknown.

6. Experiment: 3 coins are tossed. Heads and tails are recorded.

   (a) Name the 8 possible outcomes.

   (b) Suppose the coins are fair. Give the probability of 0 heads, exactly 1 head, exactly 2 heads, and 3 heads.

   (c) If the experiment is repeated 200 times, give the expected frequency of 0 heads, exactly 1 head, exactly 2 heads, and 3 heads.

7. A spinner like the one at right is spun. Assume the spinner is perfectly balanced.

   (a) Estimate the probability of landing in each numbered region.

   (b) If the spinner is spun 10 times, give the expected frequency of each number.

   (c) If the spinner is spun n times, give the expected frequency of each number.

   (d) If the spinner is spun t times, give the expected frequency for landing on an even number.

C. 8. A company has 800 employees, of whom 300 are women. If 10 promotions are made randomly, what is the expected frequency of female promotions?

C 9. Consider a high school with 200 male graduates and 225 female graduates. If the sex of a student has nothing to do with going to college and 300 students from this school go on to college, how many male graduates would you expect to go to college?

Questions for discussion

1. The blood type probabilities for U.S. Blacks are generally between those for U.S. Whites and West Africans. Why is this not surprising?

2. What blood types are there other than those given in Questions 2-5 on page 586?

Lesson 4

Unusual Events

Suppose you want to test whether a coin is fair. So you toss it 10 times. If it lands heads up all 10 times, you probably would <u>think</u> the coin is weighted in favor of heads. Why? - because the probability of a <u>fair</u> coin landing on heads 10 out of 10 times is $\frac{1}{1024}$, a very small number. But you can never be sure. 10 out of 10 <u>is</u> possible.

What if it lands on heads 8 times out of 10? Should you think it is weighted? People who use statistics often use the following procedure.

To test whether outcomes are occurring randomly:

1. Conduct an experiment. Suppose outcome Z occurs.

    Example 1: Toss a coin 10 times.
    Suppose Z = HHHTHHHHTH.

2. Let E be the event consisting of all outcomes whose deviation from the expected (assuming randomness) is as much or more than Z.

    E consists of outcomes which are at least as far from 5 heads as Z. That is, E = getting 8, 9, or 10 heads or 8, 9, or 10 tails.

3. Calculate P(E).

    It happens that in this case, P(E) = $\frac{7}{64}$ = .10...

4. If P(E) < .05, then question randomness.

 If P(E) < .01, then strongly question randomness.

Since P(E) > .05, the outcome Z does not deviate enough from 5 heads to question randomness. 8 heads is not unusual enough.

Example 2: If you get 10 heads in 10 tosses of a coin, should you question randomness?

Solution: Use the above 4 steps.

1. Z = HHHHHHHHHH = getting 10 heads

2. So E = getting 10 heads or 10 tails

3. P(E) = P(10 heads) + P(10 tails)

$$= \frac{1}{1024} + \frac{1}{1024} = \frac{2}{1024}$$

4. Since $\frac{2}{1024}$ < .01, you should strongly question either the randomness of your tossing or the fairness of the coin.

The more times an experiment is done, the more often unusual events will occur. All heads is unusual for a fair coin tossed 10 times. But if 500 people each toss a coin 10 times, then $\frac{500}{1024}$ is the expected frequency of 10 heads. So you should not be surprised if one of the 500 tossed all heads.

Similarly, it is unusual for a high temperature record to be set today. But in a year (doing the experiment 365 or 366 times), it is not unusual to have a number of high temperature records set.

589

Repeating an experiment increases the probability of the occurrence of an unusual event.

Most people consider an event unusual if its expected frequency is less than .05. An event is highly unusual if its expected frequency is less than .01.

In summary:

Expected frequency of event	Event	Should you question randomness of occurrence of outcomes?
e.f. ⩾ .05	not unusual	no
.01 ⩽ e.f. < .05	unusual	yes, but not strongly
e.f. < .01	highly unusual	strongly question, but it still could be random

Questions covering the reading

1-3. True or False.

1. It is possible to tell whether outcomes are occurring randomly.

2. Tossing a coin will yield random outcomes.

3. If you toss a coin 10 times and get 10 tails, then the coin is weighted.

4-7. Suppose you wish to test whether outcomes to an experiment are occurring randomly. You run an experiment and outcome Z occurs.

4. What is the event E for which you should calculate P(E)?

5. If $P(E) = \frac{1}{10}$, should you question randomness?

6. If P(E) = .02, should you question randomness?

7. When should you strongly question randomness?

8-10. A coin is tossed 10 times. To test randomness, for what event E should you calculate P(E) if the outcome has:

8.   8 heads   9.   9 heads   10. 10 heads

11. When is an event unusual?   12. When is an event highly unusual?

13-15. If a coin is fair, is the given event unusual? Is the given event highly unusual?

13. 5 heads or 5 tails in 5 tosses.

14. 6 heads or 6 tails in 6 tosses.

15. 7 heads or 7 tails in 7 tosses.

16. In a class of 25 students, everyone tosses a coin 10 times. Would it be unusual for one of the students to get all heads?

17. Suppose 25 students each tossed a coin 9 times. Would it be unusual for one of the students to have the coin land on the same side all 9 times?

18. Explain how a high temperature record can be considered unusual but also how it can be considered not so unusual.

Questions applying the reading

C 1-6. If a coin is tossed 15 times, here are probabilities (assuming outcomes occur randomly) of some events:

$$P(10 \text{ heads}) = \frac{3003}{32768} \quad P(11 \text{ heads}) = \frac{1365}{32768} \quad P(12 \text{ heads}) = \frac{455}{32768}$$

$$P(13 \text{ heads}) = \frac{105}{32768} \quad P(14 \text{ heads}) = \frac{15}{32768} \quad P(15 \text{ heads}) = \frac{1}{32768}$$

Should you question the fairness of a coin if you toss it 15 times and get:

1.   10 heads.   2.   11 heads.   3.   12 heads.

4.   13 heads.   5.   14 heads.   6.   15 heads.

591

7. A friend of yours gets 8 right on a 10-question true-false test and says "I guessed!" Should you believe your friend?

8-10. Suppose $P(E) = x$. Then, if the experiment is done twice,

$$P(E \text{ occurs at least once}) = 2x - x^2.$$

8. If two people each toss a fair coin 4 times, give the probability that at least one of them gets all heads.

9. If two people each toss a fair die, what is the probability that at least one of them tosses a 5?

10. If the probability of twins is 1/86, what is the probability of twins in at least one of two pregnancies?

C 11. From 1966 to 1970, there were 19 no-hitters in Major League baseball. About 1900 games are played in a season. (a) Suppose you went to a baseball game. Would it be an unusual event if you saw a no-hitter? (b) Suppose you watched 60 games a year on television. Would it be unusual for you to see a no-hitter?

12. Temperature records exist for Chicago since the Great Fire of 1871. On a particular date, the probability that a record will be broken is about $1/(y-1870)$, where y is the year. (a) What is the probability that the Chicago high temperature record for July 1st will be broken on July 1st, 1990? (b) In the entire year 1990, how many high temperature records would be expected to be broken? (c) In this question, in calculating the probability, what was assumed about the weather in Chicago?

Lesson 5

Intersection of Sets

Situation 1: A conversation between two people.

        First person: "It sure is cold today. It must be below

             freezing."

        Second person: "But it could be colder. It's not below

             zero."

        What is the temperature?

It is easiest to answer this question using graphs. The possible

(F) temperatures from what each person has said are shaded.

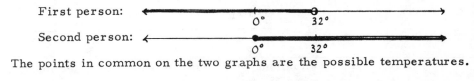

The points in common on the two graphs are the possible temperatures.

Situation 2: How many people subscribe to both <u>Time</u> and <u>Newsweek</u>?

If possible, we would examine a list of <u>Time</u> subscribers. We

would examine a list of <u>Newsweek</u> subscribers. Then we would look

for names common to both lists.

These situations fit the following pattern:

Situation	Set A	Set B	Set of elements in both sets A and B
1	set of temps. below 32°	set of temps. $\geq 0°$	temps. between 0° and 32°, including 0°
2	set of _Time_ subscribers	set of _Newsweek_ subscribers	set of people who subscribe to both _Time_ and _Newsweek_

You have probably seen the pattern before.

> **Definition:** Let A and B be two sets. The set of elements in both A and B is the intersection of the sets A and B, written $A \cap B$.

Here are more situations which use the idea of intersection.

Situation 3: Twice an integer is greater than 50. Seven times the integer is less than 100. What is the integer?

Here set A is the solution set of $2x > 50$.

$$A = \{x: 2x > 50\} = \{26, 27, 28, \ldots\}$$

Set B is the solution set of $7x < 100$.

$$B = \{x: 7x < 100\} = \{14, 13, 12, 11, \ldots\}$$

We want the set of elements that are in both A and B. But there aren't any.

$$A \cap B = \emptyset$$

There is no integer that works.

594

<u>Situation 4</u>: The word "intersection" comes from streets. Think of

streets A and B

as sets of points.

Then A ∩ B is

the shaded set of

points.

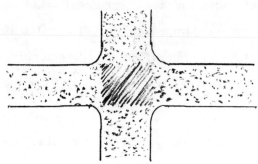

<u>Situation 5</u>: Suppose a coin and a die are tossed. There are then 12

possible outcomes.

$$\begin{array}{cccccc} \text{H1} & \text{H2} & \text{H3} & \text{H4} & \text{H5} & \text{H6} \\ \text{T1} & \text{T2} & \text{T3} & \text{T4} & \text{T5} & \text{T6} \end{array}$$

Let A = "getting heads" = {H1, H2, H3, H4, H5, H6}

B = "tossing a 4"   = {H4, T4}

Then A ∩ B = "getting heads and a 4" = {H4}

<u>Questions covering the reading</u>

1. A ∩ B is the _____ of the sets A and B.

2. If you know the elements of sets S and T, how can you find the
elements of S ∩ T?

3-6. Graph the intersection of the named sets.

3. {n: n > 5}, {n: n > 10}   4. {x: x ≤ 2}, {x: x ≥ 1.9}

5. {t: 0 < t < 6}, {t: t^2 = 4}   6. {y: y < 15}, {y: y ≥ 16}

595

7-9. Two sets are given. Describe their intersection.

7. set of people who read <u>Seventeen</u>; set of males

8. set of male chimpanzees; set of female chimpanzees

9. set of even numbers; set of prime numbers

10-12. Three coins are tossed. There are 8 outcomes in the sample space. Two events E_1 and E_2 are given. Name all elements of $E_1 \cap E_2$.

10. E_1 = 2 heads are tossed; E_2 = at least 1 head is tossed.

11. E_1 = at least 1 head is tossed; E_2 = at least 1 tail is tossed.

12. E_1 = 2 heads are tossed; E_2 = 1 tail is tossed.

13-14. A fair coin and a fair die are tossed. There are 12 outcomes as listed on p. 595. (a) Name all outcomes in $A \cap B$. (b) Calculate $P(A \cap B)$.

13. A = "getting a 4 or 6"; B = "getting heads"

14. A = "getting tails"; B = "tossing a 5"

15-20. C is a circle, m and n are lines, and P, Q, R, S, and T are points.

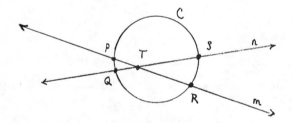

Simplify:

15. $m \cap n$ 16. $m \cap C$ 17. $n \cap C$

18. $\{P, Q, R, S, T\} \cap C$ 19. $\{P, R\} \cap n$ 20. $m \cap n \cap C$

21. Where does the name "intersection" come from?

Questions testing understanding of the reading

1-6. Two sets are called <u>mutually exclusive</u> if they have no elements in common.

1. In which of situations 1-5 in this lesson are A and B mutually exclusive?

2. If A and B are mutually exclusive, then $A \cap B$ = _____ .

3. If two lines are mutually exclusive, they are called _____ .

4. A card is picked out of a deck. Which two events are mutually exclusive?

   (a) picking a club   (b) picking a 10   (c) picking a red card

5. If A and B are mutually exclusive events, then $N(A \cap B)$ = _____ .

6. If A and B are mutually exclusive events, then $P(A \cap B)$ = _____ .

7. True or False: For any sets A and B, $A \cap B = B \cap A$.

8-11. Give an example of sets X and Y where:

8. $N(X) = 5$, $N(Y) = 4$, and $N(X \cap Y) = 2$.

9. $N(X) = 6$, $N(Y) = 3$, and $N(X \cap Y) = 4$.

10. $X \cap Y = X$ 11. $N(X \cap Y) = N(X)$

12. Give an example of 3 sets A, B, and C in which $A \cap B \cap C = \emptyset$ but $A \cap B \neq \emptyset$, $A \cap C \neq \emptyset$, and $B \cap C \neq \emptyset$.

13-15. $A \cap B$ is given. What could A and B have been?

13. $A \cap B$ = set of words beginning with q and ending in k.

14. $A \cap B$ = set of triangles with all sides of the same length.

15. $A \cap B$ = set of cities in the U.S. with population between 100,000 and 1,000,000.

Independent Events

A spinner is spun and a die is tossed. What is the probability that the spinner will land on 3 <u>and</u> the die will land on 5?

Let A = spinner lands on 3   Let B = tossing a 5

We wish to know P(A ∩ B).

For this experiment, the sample space consists of <u>pairs</u> of outcomes from the spinner and die. Since the spinner had 5 possible outcomes and the die has 6, there are 30 outcomes in the sample space.

Outcomes of (spinner, die) experiment:

(1, 1)	(1, 2)	(1, 3)	(1, 4)	(1, 5)	(1, 6)
(2, 1)	(2, 2)	(2, 3)	(2, 4)	(2, 5)	(2, 6)
(3, 1)	(3, 2)	(3, 3)	(3, 4)	(3, 5)	(3, 6)
(4, 1)	(4, 2)	(4, 3)	(4, 4)	(4, 5)	(4, 6)
(5, 1)	(5, 2)	(5, 3)	(5, 4)	(5, 5)	(5, 6)

The event A is the 3rd row. Event B is the 5th column. A ∩ B = {(3, 5)}. If the outcomes occur randomly, then the probability of getting $(3,5)$ is $\dfrac{N(A \cap B)}{N(S)}$ or $\dfrac{1}{30}$.

You might say: "Wait? Isn't there an easier way? P(spinner lands on 3) = $\dfrac{1}{5}$. P(tossing a 5) = $\dfrac{1}{6}$. Can't I just multiply the fractions to get the answer $\dfrac{1}{30}$?"

This reasoning is correct because the occurrence of event A does not affect the probability of B. Such events are called independent. The probabilities of independent events satisfy the following definition.

Definition:

> Events A and B are <u>independent</u> when and only when $P(A \cap B) = P(A) \cdot P(B)$.

Example 1: On a TV show there are 3 doors. The grand prize for the day is behind one of them. What is the probability that contestants will guess the correct door 2 days in a row?

Solution: It is realistic to think that the guesses of contestants are random and independent. (Contestants usually don't know what others have guessed.) Let A = correct guess on first day. B = correct guess on second day. Then $A \cap B$ = correct guesses on both days. Because of randomness, $P(A) = \dfrac{1}{3}$ and $P(B) = \dfrac{1}{3}$. Because of independence, $P(A \cap B) = P(A) \cdot P(B) = \dfrac{1}{3} \cdot \dfrac{1}{3} = \dfrac{1}{9}$.

Example 2: The probability of getting a sum of 7 from two fair dice

is $\frac{1}{6}$. What is the probability of getting three 7's in a

row?

Solution: Assume the tosses are independent.

P(7 on 1st toss $\cap$ 7 on 2nd toss $\cap$ 7 on 3rd toss)

= P(7 on 1st toss) • P(7 on 2nd toss) • P(7 on 3rd toss)

= $\frac{1}{6}$ • $\frac{1}{6}$ • $\frac{1}{6}$

= $(\frac{1}{6})^3$ = $\frac{1}{216}$

Example 2 illustrates what can happen if you repeat an experiment.

Repeated Experiment
Theorem:

> If an event has probability p
> and the experiment is indepen-
> dently repeated n times, then
> P(event occurs n times) = p^n.

Example 3: Suppose a person is 98% accurate on arithmetic problems.
What is the probability the person will get 10 out of 10
correct on a test?

Solution: If the problems are independent, by the repeated experiment theorem, the probability is

$(.98)^{10}$ or about .817

This is less than many people expect.

Questions covering the reading

1. When are two events A and B independent?

2-5. Consider the spinner-die experiment (as on page 598). Assume outcomes occur randomly. Let

A = spinning a 3   B = spinning a 1 or 2

C = tossing a 4   D = tossing an even

2. How many outcomes are in the sample space?

3. Calculate P(A), P(C), and P(A ∩ C). Are A and C independent?

4. Calculate P(B), P(A), and P(A ∩ B). Are A and B independent?

5. Calculate P(B), P(D), and P(B ∩ D). Are D and B independent?

6-9. Each day on a TV show, one contestant gets a chance at guessing the 1 door out of 3 from which a grand prize will appear. Calculate the probability that the program will have to give out the grand prize:

6. 2 days out of 2.   7. 3 days out of 3.

8. 4 days out of 4.   9. n days out of n.

10. State the Repeated Experiment Theorem.

C 11-14. In the past, a person has been able to type a line with no errors about 95% of the time. Assume that an error on one line does not affect the accuracy on any other line. Calculate the probability that the person will have no errors in:

11. 2 lines of typing.   12. 4 lines of typing.

13. 20 lines of typing.   14. ℓ lines of typing.

15. Why in Questions 11-14 is it necessary to assume that an error on one line does not affect accuracy on other lines? Is it reasonable to assume this?

1-4. You have reason to believe that a coin is __weighted__, with P(heads) $\approx .6$. If this is so, what is the probability of getting:

1. 2 heads in 2 tosses.   2. 3 heads in 3 tosses.

3. 4 heads in 4 tosses.   4. n heads in n tosses.

5. __Twins.__ There is a probability of about 1/86 that a birth will result in twins.

   (a) If births are independent, what is the probability that two consecutive births will each result in twins?

   (b) Actually, many more families than would be expected have two sets of twins. What do you think accounts for this?

6. A fair coin is tossed 4 times. Let E_1 = getting exactly one head in the first two tosses. Let E_2 = getting two heads exactly in the first four tosses. (a) Show that E_1 and E_2 are not independent. (Hint: Calculate $P(E_1)$, $P(E_2)$, and $P(E_1 \cap E_2)$.) (b) What does the non-independence mean?

7. You are one of 5000 people who mail for 2 tickets (maximum allowed) to a rock concert. The auditorium can seat 3000. Then you mail in for 2 tickets to a second concert where 2000 people mail for tickets. Let T_1 = getting a ticket to the first concert. Let T_2 = getting a ticket to the second concert.

   (a) Describe the event $T_1 \cap T_2$.

   (b) Calculate $P(T_1 \cap T_2)$.   (c) Are T_1 and T_2 independent?

8. In a particular year, suppose that Harvard had about 3 qualified applicants for each opening in its entering class. Yale had about 2.8 qualified applicants for each opening. If these colleges chose qualified applicants independently and at random, what was the probability of a qualified person being accepted to __both__ colleges?

9. An apprentice program has 136 applicants for 50 openings. A second program has 200 applicants for 65 openings. Suppose that each program fills its openings randomly from the applicants. If you apply to both programs, what is the probability that you will have a choice of programs?

10. Suppose that, in a particular city, it rains about 240 days a year.

   (a) If you visit this city for 1 day, what is the probability that you will get a day without rain?

   (b) Assume that rain on one day does not affect rain on a second day. If you visit the city for 2 days, what is the probability that both days will not have rain?

(c) Why is the assumption in part (b) a poor assumption to make?

11-15. A professional basketball player makes about 75% of free throws. What is the probability that this player will:

11. make 2 in a row in a key situation.

12. miss 2 in a row in a key situation.

13. make 1, then miss 1 in a key situation.

14. miss 1, then make 1 in a key situation.

15. make 3 in a row in key situations.

16-20. Answer Questions 11-15 for a player whose probability of success is p.

Lesson 7

The So-Called "Law of Averages"

In practice, we do not know probabilities. All we have is data.

Question: A coin is actually tossed 10 times. Heads occurs on the first 9 tosses. What should you expect to happen on the 10th toss?

Choices: (a) Tails is more likely.

(b) Heads and tails are equally likely.

(c) Heads is more likely.

Make an answer to the above question before reading even the next sentence. There are actually two choices which could be correct, so you have a good chance of making a correct choice.

Most people answer (a). These people argue that since the probability of heads and tails is usually equal, heads and tails must come up equally often. So the number of tails must catch up to the number of heads. So tails is more likely.

This argument is phony. The tosses can be considered as independent because the coin does not have a memory. It does not think "I came down heads up 9 times, so I must balance." It can't think. The first 9 tosses do not affect the 10th. If in fact the coin is fair, then heads and tails are equally likely. So choice (b) is correct.

But choice (c) has the data going for it. Either an unusual event

604

has occurred (for a fair coin, the probability of 9 heads in 9 tosses is $\frac{1}{512}$) or the coin is weighted in favor of heads. The data could lead one to reasonably believe that either the coin or the tossing is biased for heads and so heads should be expected on the 10th toss.

In short, you were correct if you made either choice (b) or (c). But choice (a) is incorrect. The 10th toss is independent of the other tosses.

The incorrect reasoning which leads to choice (a) is known as applying the "law of averages." The law of averages is that the longer an event doesn't occur, the greater (!) its probability of occurring. Here are three common situations in which people apply this law. In each case, the argument is false.

Situation 1: A team has lost 5 games in a row. An announcer says "They're <u>due</u> for a win."

Losing 5 games may make a team try harder to win the next game. But trying harder may not help. They may get nervous. Losses often lead to less morale and a team gets worse, not better. For these reasons, you can't tell how one game will affect the next. So without more knowledge, you should assume that each game played is independent of others. Then the probability of winning or losing is not based on past performance.

605

Situation 2: A gambler is losing. The gambler continues because he or she feels that the luck must even out.

The tragedy of many gamblers is that they believe in the law of averages. But the odds favoring winning do not change just because you are losing. The odds remain the same. (It is human nature not to apply the law of averages when you are winning.)

Situation 3: It hasn't rained in many days. So a weather forecaster feels that it's bound to rain soon.

The rain situation may be most obvious. Data collected over many years shows that dry spells can continue for long periods of time. After a dry day, sunshine is more likely than rain.

Questions covering the reading

1. What is the "law of averages?"

2. Is the "law of averages" true?

3. Suppose you toss a coin and heads occurs. Which is more likely on the next toss, heads or tails?

4. How can the law of averages lead to incorrect weather predictions?

5. How are gamblers hurt if they believe in the law of averages?

Questions testing understanding of the reading

1. Suppose you have had quizzes in a biology class three Fridays in a row. You would probably expect a quiz on the fourth Friday. According to the "Law of Averages," what should you expect?

2. A fair coin is tossed 10 times. Let A = getting 9 heads in the first 9 tosses and B = getting heads on the 10th toss. Calculate P(A), P(B), and P(A ∩ B). Are A and B independent?

3. Repeat Question 2 if the coin is weighted so that the probability of heads on any single toss is 0.65.

4. A baseball batter has been up 3 times without a hit. When she is up for the 4th time, should the announcer say, "She's due for a hit?" Explain your answer.

Questions for discussion and exploration

1. Suppose you toss a coin 100 times. How many times during the tossing will you have tossed the same number of heads as tails? (a) Guess at an answer to this question. (b) See how close your guess is by actually performing the experiment.

2. A family has 3 daughters and no sons and the mother is again pregnant. What would you tell the mother about her chances for giving birth to a son?

Chapter Summary

There are three threads of ideas intertwined in this chapter. The first thread consists of ideas about sets. Just as a set may have subsets and elements, every sample space has events and outcomes. This makes it possible to apply ideas about sets to ideas about events. The _intersection_ $A \cap B$ of two sets A and B consists of those elements in both A and B. The intersection of two events is the event where both given events occur. $N(A)$ is the number of elements in the set or event A; $P(A)$ is the probability of A.

A second thread involves repeating an experiment. If an event occurs with probability p and you repeat the experiment n times, then the _expected frequency_ of the event is np. The probability that the event occurs every time in the n repetitions is p^n. Both the calculation of expected frequency and the Repeated Trials Theorem assume that the occurrence of the event does not affect the probability of its occurring again. This is the idea behind the concept of independence. Two events A and B are independent if $P(A \cap B) = P(A) \cdot P(B)$.

608

The third thread of this chapter is an attempt to correct some misconceptions about what happens when experiments are repeated. One misconception, the "law of averages," is that later results of an experiment will make up for earlier ones. By applying this law, gamblers often destroy themselves. A second misconception concerns unusual events. An event which would be unusual if an experiment were done once may not be at all unusual if the experiment is repeated many times. Most records in weather or climate are not unusual because there are so many days on which records could be set.

Lesson 1

Introduction to Systems

A service club made $49 from a car wash. The members decided to give some of the money to the Boy Scouts, some to the Girl Scouts. If B and G stand for the number of dollars given, then

$$B + G = 49.$$

Recall that every solution to this equation is an ordered pair. The solution $(2, 47)$ means that $2 could be given to the Boy Scouts, $47 to the Girl Scouts.

There were twice as many girls as boys in this club. The girls argued that the Girl Scouts should get twice as much as the Boy Scouts. In symbols, the girls wanted

$$G = 2B.$$

How much would go to each group under the girls' plan? To answer this question, we need to find an ordered pair which works in <u>both</u> equations

$$B + G = 49$$
$$G = 2B$$

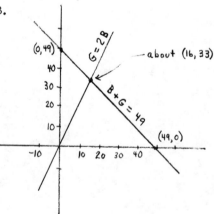

610

Each equation has a graph which is a line. The lines are graphed on p. 610. The ordered pair we want is the <u>intersection</u> of the lines. Graphing shows that the intersection is near (16, 33). Under the girls' plan, about \$16 would be given to the Boy Scouts, \$33 to the Girl Scouts.

Graphing does not always give exact solutions. Although 16 + 33 = 49, 33 is not 2•16. So (16, 33) is only close to a solution.

When you want to find all common solutions to two or more sentences, the set of sentences is called a <u>system</u>. A large brace $\{$ is sometimes used to identify a system.

$$\begin{cases} B + G = 49 \\ \qquad G = 2B \end{cases} \text{ is a system of linear equations.}$$

The solution set to a system is the intersection of the solution sets of the individual sentences. The intersection can sometimes be shown by graphing. Another method is trial and error.

Example: Solving a system by trial and error.

$$\text{Given} \begin{cases} x + y = 10 \\ x - y = 6 \end{cases}$$

The goal is to find a pair (x, y) which works in both sentences.

Solution: Because x + y = 10, try numbers which add to 10. Here

are some: 5 and 5, 4 and 6, 3 and 7, 2 and 8, 1 and 9,

0 and 10, -1 and 11, etc. Since x - y = 6, look for numbers

whose difference is 6. 2 and 8 will do.

$$\text{Notice:} \qquad \begin{cases} 8 + 2 = 10 \\ 8 - 2 = 6 \end{cases}$$

So the ordered pair (8, 2) is a solution. It is the only solu-

tion because the graph of each given equation is a line and

two different lines intersect in at most one point.

Questions covering the reading

1-3. (a) Name three ordered pair solutions to each equation. (The
order is alphabetical, as usual.) (b) By trial and error, try to find
the common solution.

1.   a + b = 21   2.   x = 3y   3.   2 = m - n

     a - b = 3   x - y = 40   0 = m + n

4-6. Graph the equations of Questions 1-3. Use a single graph for
each system.

7.   Multiple choice. In this lesson, a _system_ is:
     (a) a method of solving an equation with more than one variable.
     (b) a method of finding the solution to a set of sentences.
     (c) a set of sentences for which a common solution is wanted.

8.   Give an example of a system.

9.   The solution set of a system is the intersection of _____.

10. Name 2 ways of solving systems.

612

11. A service club has $75 which they want to give to two charities, the American Cancer Society (C) and the Heart Fund (H).

   (a) Letting C and H be the amounts given to the charities, what equation must be satisfied by (C, H)?

   (b) Give three solutions to the answer to part (a).

   (c) The graph of the solution set to the equation of part (a) is a _____.

12. Refer to the situation of Question 11. Suppose the club wants to give 2 times as much to the Cancer Society as to the Heart Fund.

   (a) From this information, what equation must be satisfied by C and H?

   (b) Give three solutions to your answer to part (a).

13. Find an ordered pair which is a solution to the equations from both Questions 11(a) and 12(a).

14. Graph the equations from both Questions 11(a) and 12(a) on the same axes.

15. The boys in the service club of this lesson decided that they did not like the girls' way of splitting up the profits from the car wash. They suggested a compromise. For every $3 given to the Boy Scouts, give $4 to the Girl Scouts. (The Girl Scouts get 4/3 as much as the Boy Scouts.)

   (a) What two equations are now satisfied by B and G?

   (b) Use either graphing or trial and error to estimate how much the Girl Scouts would get under this arrangement.

16-17. <u>Multiple choice.</u> Choose the ordered pair at right which is a solution to the system at left.

16. $\begin{cases} x - 2y = 10 \\ 3x + 4y = 40 \end{cases}$

   (a) (20, 5)   (b) (10, 0)

   (c) (12, 1)   (d) (0, 10)

17. $\begin{cases} B = 4A \\ A - B = 70 \end{cases}$

   (a) (14, 56)   (b) (56, 14)

   (c) (14, -56)   (d) (56, -14)

Questions extending the reading

1-6. Each situation leads to a system of two equations. Name the equations.

1. The sum of two numbers (call them m and n) is 72. Twice one number subtracted from the other is 60.

2. One number is three times a second number. When the first number is subtracted from the second, the difference is 1/2.

3. A will states that John Doe is to get three times as much as Mary Roe. The total amount to be given John and Mary is $11,000.

4. A sculptor wants to make two similar clay figures, one 3 times as high (so requiring 27 times the material) as the other. Six liters of clay are available.

5. On a menu, you can buy 2 eggs with sausage for $1.80. You can buy 1 egg with sausage for $1.35.

6. A suit with two pair of slacks sells for $109.50. The same suit with one pair of slacks is $89.50.

7-8. These systems are not linear systems and each has two ordered pair solutions. Finding one of the solutions answers this question. Can you find both?

7. $\begin{cases} x + y = 10 \\ xy = 21 \end{cases}$

8. $\begin{cases} A - B = 5 \\ 300 = AB \end{cases}$

Lesson 2

Solving Systems by Substitution

Graphing does not always show exact solutions. Trial and error doesn't always work. So better methods are needed for solving systems. One basic method is substitution.

You have already used substitution many times. You know the F-C relationship.

$$F = \frac{9}{5}C + 32$$

If you want the Fahrenheit equivalent of $30°$ Celsius, then you can think of solving the system

$$\begin{cases} C = 30 \\ F = \frac{9}{5}C + 32 \end{cases}$$

All you have to do is substitute 30 for C in the second equation.

$$F = \frac{9}{5} \cdot 30 + 32 = 86$$

So $(C, F) = (30, 86)$.

> When you know one coordinate of an
> ordered pair solution, you can often
> substitute to find the other coordinate.

This idea can be used to solve the Boy Scout-Girl Scout system of Lesson 1.

1.
$$\begin{cases} B + \underline{G} = 49 \\ G = 2B \end{cases}$$

The second equation indicates that G is "known." That is, G equals 2B. So 2B can be substituted for G in the top equation. The under-line shows where.

2.
$$\begin{cases} B + \underline{2B} = 49 \\ G = 2B \end{cases}$$

After substitution, the top equation has only one variable. So a value for B can be found.

3.
$$B + 2B = 49$$
$$3B = 49$$
$$B = \frac{49}{3} = 16\frac{1}{3}$$

Now B is really known. So its value, $16\frac{1}{3}$, can be used to calculate G.

4.
$$G = 2B = 2(16\tfrac{1}{3}) = 32\tfrac{2}{3}$$

Thus $(B, G) = (16\frac{1}{3}, 32\frac{2}{3})$. That is, $16\frac{1}{3}$ should be given to the Boy Scouts, $32\frac{2}{3}$ to the Girl Scouts. But there are no coins for $16\frac{1}{3}$. The club members will have to approximate.

Here is a different type of problem. It also can be solved by substitution.

Some people want to paint a room light blue.
The color can be made by mixing 2 parts
blue paint with 5 parts white paint. If it
takes 16 liters of paint for this room, how
much blue paint and how much white paint
should be purchased?

The problem has lots of words. So it is best to proceed slowly.

First identify the unknowns, the variables to be used.

Let b = amount of blue paint needed (in liters).

w = amount of white paint needed (in liters).

p = amount in one "part."

Now take the given information a little bit at a time.

Because there are 5 parts white paint: $w = 5p$

Because there are 2 parts blue paint: $b = 2p$

Because 16 liters of paint are needed: $b + w = 16$

At the right is a system of three equations. It is easily solved by

substituting 5p for w and 2p for b in the third equation.

$$2p + 5p = 16$$

Now solve for p.

$$7p = 16$$

$$p = \frac{16}{7}$$

The value of p can be used to calculate values for b and w.

$$b = 2p = 2 \cdot \frac{16}{7} = \frac{32}{7}$$

$$w = 5p = 5 \cdot \frac{16}{7} = \frac{80}{7}$$

So $\frac{32}{7}$ liters of blue paint and $\frac{80}{7}$ liters of white paint are needed.

You would probably want to buy a little more than these amounts.

Here is a fourth example of substitution.

Given:
$$\begin{cases} y = 2x - 3 \\ 3x - 2y = 10 \end{cases}$$

Substitute <u>2x - 3</u> for y in the second equation.

$$3x - 2(2x - 3) = 10$$

Now apply the distributive property to help simplify and solve.

$$3x - 4x + 6 = 10$$

$$-1 \cdot x + 6 = 10$$

$$-1 \cdot x = 4$$

$$x = -4$$

Since y = 2x - 3:

$$y = 2 \cdot -4 - 3 = -8 - 3 = -11.$$

The solution is (-4, -11).

Questions covering the reading

1. Determine (C, F).

$$\begin{cases} C = -10 \\ F = \frac{9}{5}C + 32 \end{cases}$$

2. Determine (ℓ, w).

$$\begin{cases} \ell = 40 \\ \ell w = 100 \end{cases}$$

3. Determine (x, y).

$$\begin{cases} x - 4y = 7 \\ \quad y = 2 \end{cases}$$

4. Determine (u, v).

$$\begin{cases} \frac{1}{2} = u \\ 3u - v = 6 \end{cases}$$

5. What real problem might be solved by doing Question 1?

6. What real problem might be solved by doing Question 2?

7-10. Use substitution to help solve each system.

7. $\begin{cases} x + y = 14 \\ \quad x = 6y \end{cases}$
 8. $\begin{cases} 2m - 3n = -15 \\ \quad\quad m = 4n \end{cases}$

9. $\begin{cases} \quad b = 3a \\ 60 = a - b \end{cases}$
 10. $\begin{cases} \quad\quad y = x + 2 \\ 2x + 4y = 29 \end{cases}$

11-14. (a) Write the two equations for a system which would help answer the question. (b) Answer the question by solving the system.

11. A club wants to give 3 times as much to charity as they spend on a dance. If there is $375 to be split up, how much will the club have for the dance?

12. A football stadium seats 20,000 people. If the home team gets 4 times as many tickets as the visitors, how many tickets does each team get?

13. A painter needs 12 liters of a light green paint which is 3 parts white and 2 parts green. How many liters of each color should be used?

14. A certain shade of maroon is a mixture of 3 parts dark brown and 5 parts red. How many bottles of each color are needed to make 12 bottles of maroon?

Questions extending the reading

1. Given the system $\begin{cases} y = 2x + 3 \\ y = 3x + 20 \end{cases}$

   a student substituted <u>2x + 3</u> for y in the second equation. This

   resulted in

   $$2x + 3 = 3x + 20.$$

   (a) Is this correct procedure?

   (b) If the procedure is correct, finish solving the system. If the procedure is not correct, tell what is wrong with it.

2. Given the system

$$\begin{cases} w = -9z \\ z - 2w = 323 \end{cases}$$

a student substituted $-9z$ for w in the second equation. This resulted in $z - 18z = 323$.

(a) Is this correct procedure?

(b) If the procedure is correct, finish solving the problem. If the procedure is not correct, tell what is wrong with it.

3. According to The Consumer's Handbook (edited by Paul Fargis, published by Hawthorn Books, 1974), a good oiled furniture finish contains 2 parts boiled linseed oil and 1 part turpentine. If you need 4 ounces of this furniture finish, how many ounces of turpentine are needed?

4-5. After staining furniture, The Consumer's Handbook urges that you protect the color by sealing the stain. "For a homemade sealer, mix 1 part shellac and 5 parts denatured alcohol." (Quote from page 90.)

4. If you need 160 grams of sealer, how many grams of shellac are needed?

5. If you need 160 grams of sealer, how many grams of denatured alcohol are needed?

6. To make "easy fudge frosting," the recipe calls for

    1 part hot light cream
    3 parts chocolate chips
    3 parts miniature marshmallows

If you are making a large cake for which 5 cups of frosting is needed, how much of each ingredient will you need?

7. For a "pink Lassie," put the following ingredients into a blender and blend until smooth.

    1 part orange juice
    4 parts cranberry juice
    4 parts vanilla ice cream

If you want to fill a 10-liter punch bowl with this drink, how much of each ingredient will you need?

8. Starting new plants (particularly ferns), the following soil mixture is recommended:

> 2 parts garden soil pasteurized in an oven
> 2 parts peat moss or leaf mold
> 1 part sharp sand

How much is needed of each ingredient to fill a 2-liter clay pot?

9-10. In the news, profits of a company from this year (T) are often compared with those from last year (L). Determine T and L from the given information.

9. This year the company made $200,000 more than last. The profits were up 25%.

C 10. Profits were down 8%. The company made $6.2 million more last year than this year.

Lesson 3

Solving Systems Using Addition or Subtraction

You have learned the following ways of solving systems.

    1) Graphing

    2) Trial and error

    3) Substitution

Another way to solve systems is the <u>addition</u> method. It is sometimes the easiest way.

Example 1 (Addition method): Find all solutions to the system

$$\begin{cases} 2x + 3y = 12 \\ 9x - 3y = 10 \end{cases}$$

Solution: <u>Add</u> the left sides and the right sides.

$$11x = 22$$

Now it is easy. $x = 2$

Substituting 2 for x in the top equation

$$2 \cdot 2 + 3y = 12$$

Solving $y = \dfrac{8}{3}$ Thus $(x, y) = (2, \dfrac{8}{3})$.

This is the only possible solution and it does check in both equations.

Addition works because if two numbers are equal and two other numbers are equal then their sums

$\dfrac{1}{2} = .5$ or $a = b$

$\dfrac{1}{4} = .25$ or $c = d$

are also equal. (This is the Addition Property of Equations first
studied in Chapter 3.) The sum is
not changed by writing the numbers
differently.

$$\frac{1}{2} + \frac{1}{4} = .5 + .25 \text{ or } a + c = b + d$$

$$\frac{3}{4} = .75$$

Similarly, you could <u>subtract</u> and get equal answers.

$$\frac{1}{2} - \frac{1}{4} = .5 - .25$$

Subtraction can be helpful when addition is not.

Example 2 (Subtraction method): Solve the system

$$\begin{cases} -11 = 6m + b \\ 3 = 4m + b \end{cases}$$

Solution: Subtract the second equation from the first.

$$-14 = 2m$$

$$-7 = m$$

To find b, substitute -7 for m in either equation. We
pick the first

$$-11 = 6 \cdot -7 + 6$$

$$31 = b$$

So $(m, b) = (-7, 31)$.

Addition helps in Example 1 because $3y + -3y = 0$. Subtraction
helps in Example 2 because $b - b = 0$. The zero result means that
the resulting equation has only one variable in it. This goal (an
equation with only one variable) determines whether you add or sub-

tract. However, it is possible that neither addition or subtraction
will work.

$$\begin{cases} A - 5B = 2 \\ 2A - 9B = 6 \end{cases} \qquad\qquad \begin{cases} A - 5B = 2 \\ 2A - 9B = 6 \end{cases}$$

Add:　　　　$3A - 14B = 8$　　　Subtract:　　　$-A + 4B = -4$

Neither resulting equation is easy to solve. A better method is discussed in Lesson 5.

Questions covering the reading

1. Name 4 ways of solving systems.

2-7. (a) Which method, addition or subtraction, could help solve the
given system. (b) Use the method to solve the system.

2. $3x + 8y = 2$ 　　　3. $2x + .3y = 1$ 　　　4. $a + b = 11$
　 $3x - 4y = 6$ 　　　　　$12x - .3y = 44$ 　　　　$a - b = 4$

5. 　$11 = 3m + b$ 　　6. $\frac{1}{2}x + 6 = 3y$ 　　7. $y = 5x + 2$
　　$-11 = 4m + b$ 　　　　　　　　　　　　　　$y = 3x + 2$

　　　　　　　　　　　　　$\frac{1}{2}x + y = 2$

8-13. Solve each system using either addition or subtraction.

8. $2x - 3y = 5$ 　　　9. $6a + 2c = 200$ 　　10. $9y + 4x = 2$
　 $5x - 3y = 11$ 　　　　$9a - 2c = 25$ 　　　　$9y - 4x = 7$

11. $u = 3v + 1$ 　　12. $-m + 4n = 1$ 　　13. $6x - 4y = 11$
　　$u = 9v - 2$ 　　　　$-m + 6n = -1$ 　　　$2x - 4y = 23$

14. Multiple choice. Upon which idea is the addition method based?

　　(a) If $a = b$ and $c = d$, then $a + b = c + d$.

　　(b) If $a = b$ and $c = d$, then $a + c = b + d$.

　　(c) If $a = b$ and $c = d$, then $a - c = b - d$.

15. Give an example of a system where neither addition nor subtraction helps to solve it.

1-4. **Equations for lines.** As you know, every non-vertical line has
an equation of the form

$$y = mx + b.$$

If (6, -11) and (4, 3) are on the line, then

$$-11 = 6m + b$$

$$\text{and} \quad 3 = 4m + b$$

In Example 2 of this lesson, p. 623, m and b were found to be -7
and 31. So an equation for the line through (6, -11) and (4, 3) is
$y = -7x + 31$. Use this idea to find an equation for the line containing:

1. (32, 0) and (212, 100)   2. (-4, -9) and (-2, 20)

3. (11, -5) and (20, 4)   4. (12, 9) and (6, 5)

5-6. **Two puzzles.**

5. Find two numbers whose sum is -1 and whose difference is 5.

6. Find two numbers whose sum is 4386 and whose difference is 2822.

7. Two eggs with bacon cost $1.45. One egg with bacon costs $.90.
    At these rates, what should bacon alone cost?

8. An ice cream cone with two scoops costs $.85. The same cone
    with one scoop costs $.50. What should a cone alone cost?

9. Use subtraction to find x when

$$x^2 + 2x - 6 = 29$$

$$\text{and} \quad x^2 - 4x - 2 = 3$$

10. If $d = .\overline{39}$, then $100d = 39.\overline{39}$. Use subtraction to find d.
    (This idea can be used to calculate a fraction equal to any
    repeating decimal.)

How Many Oranges Are in This Display?

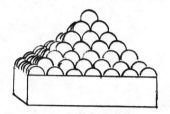

A display of oranges can be stacked in the following way: 1 orange is in the top level, 4 oranges in the 2nd level, 9 in the 3rd level, 16 in the 4th level, etc. The diagram at right shows the 16 oranges of the 4th row. The 9 oranges of the 3rd row go above the 9 dots.

So there are in all:

in the top row:	1 orange
in the top 2 rows:	5 oranges
in the top 3 rows:	14 oranges
in the top 4 rows:	30 oranges
⋮	⋮

How many oranges are needed for a display with n rows?

To answer this question, think of 1, 5, 14, 30, ... as a sequence. A formula for the nth term of this sequence is desired. To find this formula, it helps to examine other formulas for sequences.

Example 1: A polynomial formula of degree 1.

$$s_n = 6n + 5$$

The sequence is found by substituting 1 for n, then 2 for n, then 3, etc.

$$s_1 \quad s_2 \quad s_3 \quad s_4 \quad s_5 \quad \cdots$$

$$= 11 \quad 17 \quad 23 \quad 29 \quad 35 \quad \cdots$$

Now take differences of consecutive terms (right minus left).

$$6 \quad 6 \quad 6 \quad 6 \quad \cdots$$

The differences are all equal.

Example 2: A polynomial formula of degree 2.

$$s_n = 2n^2 - 9n + 6$$

Again the sequence is found by substitution.

$$s_1 \quad s_2 \quad s_3 \quad s_4 \quad s_5 \quad s_6 \quad \cdots$$

$$= -1 \quad -4 \quad -3 \quad 2 \quad 11 \quad 24 \quad \cdots$$

Now take differences of consecutive terms. Right minus left.

$$-3 \quad 1 \quad 5 \quad 9 \quad 13$$

Take differences a second time. This time the differences are all equal.

$$4 \quad 4 \quad 4 \quad 4$$

You may be able to guess what happens when the idea is used

with a polynomial formula of degree 3. The third time differences are taken, they are all equal.

Example 3: A polynomial formula of degree 3.

$$s_n = \frac{1}{2}n^3 - \frac{1}{2}n^2 + 3n - 9$$

Again it is first necessary to find the sequence.

	s_1	s_2	s_3	s_4	s_5	s_6	s_7	...
=	-6	-1	9	27	56	99	159	...

Differences once:	5	10	18	29	43	60	
twice:	5	8	11	14	17		
thrice:	3	3	3	3			

The general pattern is not difficult to state.

<u>Polynomial Pattern Theorem</u>:

> Given a sequence, take differences of consecutive terms. If the differences are equal the nth time this is done, then s_n is a polynomial of degree n.

Now look back at the problem with the oranges. Let s_n be the number of oranges in the top n rows. By counting:

$$s_1 \quad s_2 \quad s_3 \quad s_4 \quad s_5 \quad s_6 \quad \cdots$$

$$= 1 \quad 5 \quad 14 \quad 30 \quad 55 \quad 91 \quad \cdots$$

1st time: 4   9   16   25   36

2nd time: 5   7   9   11

3rd time: 2   2   2

The third time the differences are equal. By the Polynomial Pattern Theorem, there is evidence to believe that s_n is a polynomial of degree 3. That is,

$$s_n = an^3 + bn^2 + cn + d$$

where we don't know (but would like to know) a, b, c, or d. Now substitute for n.

Let n = 1:   1 = a1 + b1 + c1 + d

    n = 2:   5 = a8 + b4 + c2 + d

    n = 3:   14 = a27 + b9 + c3 + d

    n = 4:   30 = a64 + b16 + c4 + d

This gives a system of 4 equations with 4 variables. It looks hard to solve. But b16 is 16b, etc. And these systems are always easy to solve using subtraction. Here's how you can organize the work. (Always subtract an equation from the one just above it.)

$$\begin{cases} 64a + 16b + 4c + d = 30 \\ 27a + 9b + 3c + d = 14 \\ 8a + 4b + 2c + d = 5 \\ a + b + c + d = 1 \end{cases}$$

$$37a + 7b + c = 16$$
$$19a + 5b + c = 9$$
$$7a + 3b + c = 4$$

$$18a + 2b = 7$$
$$12a + 2b = 5$$

$$6a = 2$$

629

From the equation $6a = 2$, $a = \frac{1}{3}$. By substitution, $b = \frac{1}{2}$. Then a and b together help find out that $c = \frac{1}{6}$. Using $a + b + c + d = 1$, $d = 0$. Thus

$$s_n = \frac{1}{3}n^3 + \frac{1}{2}n^2 + \frac{1}{6}n$$

For example, if there are 10 rows of oranges, there will be a total of s_{10} oranges in the display, where

$$s_{10} = \frac{1}{3} \cdot 10^3 + \frac{1}{2} \cdot 10^2 + \frac{1}{6} \cdot 10$$

$$= \frac{1000}{3} + \frac{100}{2} + \frac{10}{6}$$

$$= 384$$

The Polynomial Pattern Theorem describes one of the few ways to find formulas for patterns. While it is not easy to solve the system to get the coefficients in the formula, computers can do this very quickly.

<u>Questions covering the reading</u>

1. What is the Polynomial Pattern Theorem?

2-5. A sequence is given. (a) Find the differences of consecutive terms. (b) Do this until all differences are equal. (c) Each sequence has a polynomial formula. Of what degree is the polynomial?

2. 15 13 11 9 7 5 ...

3. 1 3 6 10 15 21 ...

4. 10 5 2 1 2 5 10 ...

5. 1 5 14 30 55 91 140 ...

6-9. Solve each system.

6. $9x + 3y + z = 5$
    $4x + 2y + z = 2$
    $x + y + z = 1$

7. $d + e + f = -2$
    $4d + 2e + f = 7$
    $9d + 3e + f = 5$

8. $a + b + c + d = 1$
    $8a + 4b + 2c + d = 5$
    $27a + 9b + 3c + d = 12$
    $64a + 16b + 4c + d = 22$

9. $64w + 16x + 4y + z = 85$
    $27w + 9x + 3y + z = 40$
    $8w + 4x + 2y + z = 15$
    $w + x + y + z = 4$

10. A sequence begins 2, 5, 10, ... and has a formula of the form $s_n = an^2 + bn + c$. What three equations are satisfied by a, b, and c?

11. A sequence begins 11, 10, 5, 2, ... and has a formula of the form $s_n = an^3 + bn^2 + cn + d$. What four equations are satisfied by a, b, c, and d?

12-15. Find formulas for the sequences of Questions 2-5.

Questions applying the reading

1. Some displays of oranges allow no room for "inside" oranges. The rows look as follows (as seen from the top).

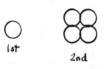

How many oranges will be in the first n rows?

C 2. A medical school has 60 first-year students, 55 second-year students, and 52 third-year students. Find a 2nd degree formula relating the year and number of students.

C 3. A restaurant charges $2.25 for a 7" pizza, $3.50 for a 10" pizza, and $5.25 for a 13" pizza. Find a 2nd degree formula connecting size and price. According to your formula: (a) What might the restaurant charge for a 16" pizza? (b) What might the restaurant charge for no pizza at all?

4. How many diagonals does a polygon have? Find a polynomial formula for the number of diagonals of an n-sided polygon using the given information. (Hint: $s_3 = 0$, $s_4 = 2$, etc.)

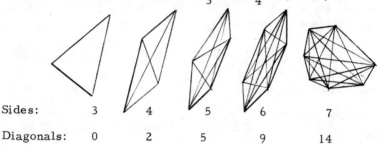

Sides: 3   4   5   6   7

Diagonals: 0   2   5   9   14

Questions for discussion and exploration

C 1. Based on tests made by the Bureau of Public Roads, here are the distances it takes to stop in minimum time under emergency conditions. Reaction time is considered to be 0.75 second. (Source: American Automobile Association; quoted in The Man-Made World, p. 15.)

mph	stopping distance
10	19 feet
20	42
30	73
40	116
50	173
60	248
70	343

(a) Graph the data on an accurate graph.

(b) Consider the stopping distances as a sequence. Take differences until they are nearly constant.

(c) Find a 3rd degree polynomial formula which fits 10, 20, 30, and 40 mph stopping values exactly. How good is the formula for 70 mph?

(d) Find a 3rd degree polynomial formula which fits 10, 30, 50, for 70 mph stopping values exactly. How good is the formula for 20 mph?

(e) What might a formula for stopping distances be used for?

2. Consider this "doubling" sequence.

$$3, \quad 6, \quad 12, \quad 24, \quad 48, \quad \ldots$$

Why can there be no polynomial formula for this sequence?

Lesson 5

Multiplying to Solve Systems

Every line has an equation of the form

$$Ax + By = C$$

You have learned how to find intersections of lines by graphing, substitution, trial and error, addition, or subtraction. But what about the following lines?

Example 1:
$$\begin{cases} 5x + 8y = 21 \\ x - 2y = -3 \end{cases}$$

Here addition or subtraction does not help. But multiply the bottom equation by 5. This makes both equations have a "5x" term.

$$\begin{cases} 5x + 8y = 21 \\ \underline{5x - 10y = -15} \end{cases}$$

Subtracting, $\qquad 18y = 36$

$\qquad\qquad y = 2$ and by substitution, $x = 1$.

The lines intersect at $(1, 2)$.

In Example 1, one multiplication was needed. In the next example, two multiplications are used. This process will always work.

634

Example 2: Find the intersection of the lines with equations

$$3x + 9y = 2$$

and $$5x - 10y = 6.$$

Solution: Neither addition nor subtraction will help. So multiply the top equation by 5, the bottom by 3. (This gives 15x in each equation.)

$$\begin{cases} 15x + 45y = 10 \\ \underline{15x - 30y = 18} \end{cases}$$

Subtract: $$75y = -8$$

$$y = -\frac{8}{75}$$

To find x, substitute $\frac{-8}{75}$ for y in either given equation.

$$3x + 9 \cdot -\frac{8}{75} = 2$$

$$3x = 2 + \frac{72}{75}$$

$$3x = \frac{222}{75}$$

$$x = \frac{74}{75}$$

So the point of intersection is $(\frac{24}{75}, \frac{-8}{75})$.

Problems with complicated mixtures can be easily solved using these ideas.

Example 3: A large ping-pong set contains 4 paddles and 6 ping-pong balls. A small set contains 2 paddles and 1 ball. A store owner receives a "broken shipment" of 100 paddles and 110 balls. Can she divide these evenly into big and small sets?

Solution: Let L be the number of large sets. Let S be the number of small sets. Keep track of paddles:

4L (in large sets) and 2S (in small sets) add to 100.

Keep track of ping-pong balls:

6L (in large sets) and 1S (in small sets) add to 110.

This gives the two equations.

$$\begin{cases} 4L + 2S = 100 \\ 6L + S = 110 \end{cases}$$

Solve as before.

$$M_2: \quad \begin{cases} 4L + 2S = 100. \\ \underline{12L + 2S = 220} \end{cases}$$

Subtract: $\qquad -8L = -120$

$$L = 15$$

$$S = 10$$

15 large sets and 10 small sets will do it.

<u>Questions covering the reading</u>

1-4. (a) Multiply one of the equations by a number which makes it possible to add or subtract to solve the system. (b) Solve the system.

1. $5x + y = 30$
    $3x - 4y = 41$

2. $6u - 5v = 2$
    $12u + 7v = 5$

3. $.05x + .06y = 18$
    $x + 2y = 400$

4. $3a - 2b = 20$
    $9a - 4b = 40$

5-8. (a) Multiply each equation by a number so that you could add or subtract to solve the system. (b) Solve the system.

5. $6m - 7n = 6$
   $7m - 8n = 15$

6. $-14 = 2A + 5B$
   $6 = 5A + 12B$

7. $7x - 3y = 200$
   $17x + 7y = 300$

8. $-3c + 2d = 8$
   $2c + 3d = 3$

9-14. Refer to Example 3 of this lesson.

9. What do L and S stand for?

10. 4L is the number of _____ in the _____ sets.

11. 2S is the number of _____ in the _____ sets.

12. 6L is the number of _____ in the _____ sets.

13. S is the number of _____ in the _____ sets.

14. Suppose the broken shipment had contained 60 paddles and 38 balls. Could the owner divide them evenly into big and small sets? (Hint: What system should be solved to answer this question?)

Questions applying the reading

1-6. Solve the system.

1. $3y - 4x = 2$
   $y + 8x = 6$

2. $3a = 2b + 5$
   $a - 4b = 6$

3. $100 = 3m - 2n$
   $60 = 4m - 3n$

4. $\frac{1}{2}x + y = 6$
   $x - \frac{1}{3}y = 12$

5. $2t + u = 46$
   $8t - 4u = 20$

6. $2v - 3w = 10$
   $3v - 4w = 12$

7. A marching band has 52 in the band and 24 in the pompon squad. They wish to form hexagons and squares like those diagrammed below.

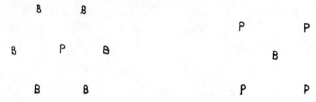

hexagon, pompon   square, band member
person in center   in center

Can it be done with no people left over? (Hint: Calculate how many band members are needed for h hexagons and s squares. Then do the same for the pompon squad.)

8. Repeat Question 7 if two more people join the band.

9-14. <u>Harder mixtures</u>. (a) Translate the given situation into two equations. (b) Solve the resulting system. Part (a) is done for Question 9 to provide an example.

9. x liters of a 60% alcohol solution are mixed with y liters of an 80% solution to get 20 liters of a 75% alcohol solution.

Part (a):   x + y = 20   (This equation tells about liters.)

            .60x + .80y = .75(20)   (This equation tells about alcohol.)

Part (b) is left for you to do.

10. From a soil analysis, it is determined that a lawn needs 50 pounds of fertilizer which is 20% nitrogen. How can this fertilizer be made from two fertilizers of 15% and 24% nitrogen?

11. How much candy worth $1.79 a pound should be mixed with candy worth $1.19 a pound to get 1 pound of candy which is worth $1.39 a pound?

12. An auditorium seats 2200 people. How many tickets should go for $3 and how many for $4 in order to get a total income of $7500 if the auditorium is full?

13. How much alcohol (a 100% solution) should be mixed with a 60% alcohol solution to get 2 liters of a 75% solution?

14. How many stamps with a catalog value of 2¢ each should be mixed with stamps of a catalog value of 10¢ each to get a mixture of 500 stamps with a catalog value of $20.00?

Summary Questions

1-6. Each question relates to the information given in Questions 1-6 on page 614. Solve a system to answer the question.

1. Find the numbers.   2.   Find both numbers.

3. How much will Mary get?

4. How much clay should be used in each figure?

5. How much should one egg cost?

6. How much should the suit jacket alone cost?

Parallel or Coincident Lines

When two lines intersect, the point of intersection can be found by solving a system.

But what happens if the lines don't intersect?

Line m has equation

$$2x + 3y = -6.$$

Line n has equation

$$4x + 6y = 24.$$

(You can check these by substituting in the points.)

Solving the system will help to find any point of intersection.

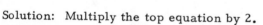

$$\begin{cases} 2x + 3y = -6 \\ 4x + 6y = 24 \end{cases}$$

Solution: Multiply the top equation by 2.

$$\begin{cases} 4x + 6y = -12 \\ 4x + 6y = 24 \end{cases}$$

The number <u>4x + 6y</u> cannot be both -12 and 24 at the same time. So the system has no solution. The solution set is ∅. If you subtracted you would get

$$0 = -36$$

another impossibility.

640

What if you picked two equations for the same line?

Example 2: Two sentences for the same line.

$$\begin{cases} 8x - 7y = 3 \\ 16x - 14y = 6 \end{cases}$$

Solution: Multiplying the top equation by 2, you get

$$\begin{cases} 16x - 14y = 6 \\ 16x - 14y = 6 \end{cases}$$

The equations are identical. Any ordered pair satisfying

the top equation will satisfy the bottom equation. So there

are infinitely many solutions. <u>The lines are coincident</u>.

These are the only possibilities, because two lines either inter-

sect in one point, do not intersect, or are coincident.

<u>Questions covering the reading</u>

1.   Multiple choice: Two lines <u>cannot</u> intersect in
     (a) exactly one point   (b) no points
     (c) exactly two points   (d) infinitely many points

2.   Give an example of equations for two parallel lines.

3.   Give an example of two different equations for the same line.

4-7.   Determine whether the lines are parallel or coincident.

4. $2x - 3y = 11$ 5. $6 = 3A + B$
     $8x - 12y = 11$ $18 = 9A + 3B$

6. $x - y = 5$ 7. $9t + 10 = 6u$
     $y - x = -5$ $15u - 20t = 5$

641

1-4. Determine whether the lines are parallel, coincident, or intersect.

1. $y = 3x + 2$
    $y = 5x + 2$

2. $2x - 5y = 6$
    $10y - 4x = 3$

3. $x + y = 10$
    $-x - y = -10$

4. $a - 3b = 2$
    $a - 4b = 2$

5-6. Could the given situation have happened?

5. 850 cheap and 250 expensive tickets were sold to a concert with total receipts of \$2175. At the same prices, 1020 cheap tickets and 300 expensive tickets netted \$2600.

6. 16 pencils and 5 erasers were bought for \$8.00; 32 pencils and 10 erasers were bought for \$16.00.

Lesson 7

Half-planes

Situation 1: Going to college, a girl budgets $100 for slacks and blouses. Slacks cost $18.00 a pair and blouses cost $8.40 each. What options does she have in buying?

You could answer this question using trial and error. However, it's much easier to use graphing.

Let s = number of pairs of slacks bought.

Total cost of slacks = 18.00s

Let b = number of blouses bought.

Total cost of blouses = 8.40b

Since the total cost cannot be greater than 100:

$$18.00s + 8.40b \leq 100$$

There are many ordered pairs which satisfy this inequality. The best way to find them all is to graph all solutions. Here is how it is done.

1. Graph the line $18.00s + 8.40b = 100$. Easy points to find are those on the axes: $(0, \frac{100}{8.40})$ and $(\frac{100}{18}, 0)$.

2. All points which satisfy $18.00s + 8.40b < 100$ lie on one side of this line. Since $(0, 0)$ works, that side must be the one containing $(0, 0)$.

3. Now s and b are positive integers or zero. So only the dotted points are possible. There are 42 points, so there are 42 options in buying. One option is circled and indicated by A. A = (3,5) and stands for buying 3 pairs of slacks and 5 blouses.

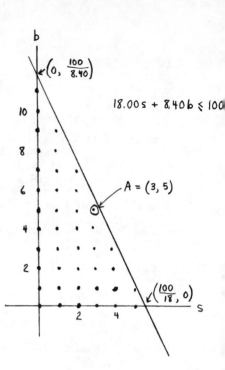

$18.00s + 8.40b \leq 100$

$A = (3, 5)$

There's another side to this situation and another side to the line. The other side shows when the store will take in <u>more</u> than $100. For example, point P indicates selling 4 pair of slacks and 7 blouses. The store will get

$18.00(4) + 8.40(7)$

or $130.80 from these sales.

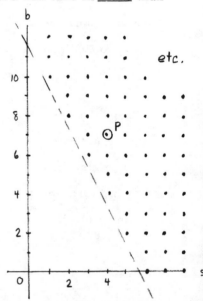

etc.

644

The general idea behind this example is quite easy. You already
know that the set of ordered pairs (x, y) satisfying

$$Ax + By = C$$

is a line. There are two other possible ways to relate $Ax + By$ to C.

$$Ax + By < C$$

$$Ax + By > C$$

Each of these inequalities corresponds to one side of the line. A side
of a line is called a half-plane.

Example: Graph all solutions to
$y > \frac{x}{2} - 1$.

Solution: The graph of $y = \frac{x}{2} - 1$
is the line with slope $\frac{1}{2}$,
y-intercept -1. Two
points on the line are
$(0, -1)$ and $(1, -\frac{1}{2})$. The
desired graph is one side
of that line. Which side?
Try $(0, 0)$ in the inequality.
It works. So $(0, 0)$ must be in the graph. The graph is

the shaded half-plane of the line. It does not include the

line so the line is dashed.

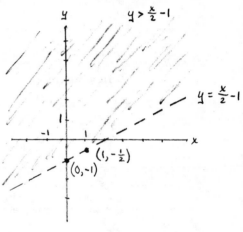

645

Situation 2: The American Society for Testing Materials classifies different types of solder (pronounced "sodder," a material heated and used to unite metals). Type 1A solder is 50% tin, 50% lead. Type 4A solder is 37.5% tin, 62.5% lead.

Suppose you wish to make these types of solder and you have plenty of tin but only 100 grams of lead. How many grams of each type can you make?

Solution: Let x = amount of type 1A solder made. Clearly $x \geq 0$.

Let y = amount of type 4A solder made. $y \geq 0$.

Then $.50x$ = amount of lead used in making type 1A solder.

$.625y$ = amount of lead used in making type 4A solder.

Since you have only 100 grams of lead in all, we know the following about x and y.

$$.50x + .625y \leq 100$$

$$x \geq 0$$

$$y \geq 0$$

The points satisfying $.50x + .625y \leq 100$ are the half-plane below the oblique line. The points with $x \geq 0$ are in the half-plane to the right of the y-axis. The points with $y \geq 0$ are in the half-plane above the x-axis. The intersection of the three half-planes contains points which work. For instance, $(90, 70)$ is in the intersection. So you could

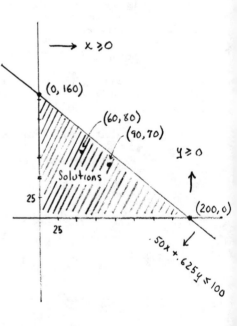

646

make 90 grams of type 1A and 70 grams of type 4A. You could also

make 60 grams of type 1A and 80 grams of type 4A because (60, 80)

is in the intersection.

Questions covering the reading

1-4. Suppose Maria has money for s slacks and b blouses. Slacks cost \$18 a pair and blouses cost \$8.40. Match each phrase at the left with the mathematical description at right.

1. total amount paid for slacks   (a) \$18s   (b) 8.40b

2. total amount paid for blouses   (c) 18s + 8.40b > 80

3. Maria spent less than \$80.   (d) 18s + 8.40b = 80

4. Maria spent more than \$80.   (e) 18s + 8.40b < 80

5-8. Graphed at right is the solution set to $4x + 3y = 24$. Describe the graph of the solution set to:

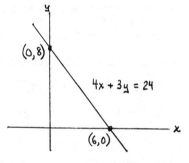

5. $4x + 3y < 24$

6. $4x + 3y \leq 24$

7. $4x + 3y \geq 24$

8. $4x + 3y > 24$

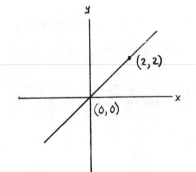

9-12. Graphed at left is the line with equation $x = y$. Describe the graph of all solutions to:

9. $x < y$ 10. $x > y$

11. $x \leq y$ 12. $x \geq y$

13-16. Match each inequality at left with its description at right.

13. x < 0   (a) half-plane above x-axis

14. y > 0   (b) half-plane below x-axis

15. y < 0   (c) half-plane right of y-axis

16. x > 0   (d) half-plane left of y-axis

17-24. Examine Situation 2 on page 646. Tell what each expression stands for.

17. .50x   18. .625y   19. .50x + .625y

20. x ⩾ 0   21. y ⩾ 0   22. .50x + .625y ⩽ 100

23. (100, 80) is a solution to .50x + .625y ⩽ 100.

24. (120, 70) is not a solution to .50x + .625y ⩽ 100.

25. What organization classifies various types of solder?

Questions testing understanding of the reading

1-9. Graph the solution set to each sentence.

1. x + y > 1   2. 3x - 2y ⩽ 6   3. a + 5b ⩾ 15

4. x + y < 9   5. 5m - 2n < 100   6. 9x + 4y ⩾ 2

7. x + 0·y > -3   8. x > -3   9. y < 5

10-11. Graph the set of points (x, y) which satisfy each situation.

10. A person wants to buy x pencils at 5¢ each and y erasers at
    15¢ each and spend less than 60¢.

11. A car dealer wants to sell x cars for about $3000 each and
    y cars for about $4500 each and make more than $27000.

12-17. <u>Accurately</u> graph the set of points which satisfy each situation.

12. <u>Basketball</u>. A team scores 2 points for a basket, 1 point for a free throw. How many baskets (b) and free throws (f) will get you over 20 points?

13. <u>Typing</u>. It takes a good typist about 10 minutes to type a letter of moderate length and about 8 minutes to type a normal double-spaced page. About how many letters (ℓ) and how many pages (p) will keep a typist busy for 6 hours?

14. <u>Hockey</u>. A hockey team figures it needs 60 points to make the playoffs. A win is worth 2 points and a tie is worth 1 point. How many wins (w) and ties (t) will get a team into the playoffs?

15. <u>Elevators</u>. Many elevators have a capacity of 1000 kg. If a child averages 40 kg and an adult 70 kg, how many children (c) and adults (a) will cause this kind of elevator to be overloaded?

16. <u>Driving</u>. Every kilometer of city driving uses 1/5 of a liter of gas. Every kilometer of country driving uses 1/6 of a liter. Assuming this is true, how far can a car go on 60 liters of gas?

17. <u>Diets</u>. One ounce of hamburger has about 80 calories and 1 "French fry" has about 14 calories. How much hamburger and how many French fries can you have if you wish to stay below 400 calories for a meal?

18-20. Graph each sentence. Then identify the set of ordered pairs which satisfy the system. What does the intersection look like?

18. $\begin{cases} x + 2y \geqslant 5 \\ x + 2y \leqslant 8 \end{cases}$

19. $\begin{cases} x \geqslant 0 \\ y \geqslant 0 \\ y \geqslant \frac{1}{2}x \\ y \leqslant x \end{cases}$

20. $\begin{cases} x + y > 1 \\ x < 1 \\ y < 1 \end{cases}$

Chapter Summary

A <u>system</u> is a set of sentences in which it is desired to find all common solutions. The solution set to a system is the intersection of the solution sets to the sentences. There are four major ways of solving systems: (1) graphing, (2) trial and error, (3) substitution, and (4) multiplication followed by addition or subtraction.

Graphing enables solutions to systems to be pictured. This method is particularly useful when the system involves inequalities, because it is easier to picture a large number of solutions than to list them all. The simplest inequalities are of the form $ax + by < c$; the graph of each is a <u>half-plane</u>. Graphing also helps explain why certain linear systems have no solution.

Trial and error is a fundamental way of checking any sentence-solving. For simple systems it is useful.

Substitution is a useful method when a value for one variable is given, or when one variable is known in terms of a second variable.

The most widely applicable of the methods is multiplication followed by addition or subtraction. When the given system is of the

form
$$\begin{cases} ax + by = c \\ dx + ey = f, \end{cases}$$
the idea is to multiply the top or bottom equations by numbers so that when the resulting equations are added, a new equation in only one variable results.

The variety of problems which can be solved by systems is endless. Some of the major types of problems found in this chapter are splitting up moneys in various ways, working with mixtures in which "parts" or percentages are given, finding polynomial formulas for sequences, and determining prices or numbers of articles which can be purchased.

CHAPTER 15

QUADRATIC EQUATIONS

Lesson 1

Quadratic Expressions

A $\underline{\text{quadratic equation}}$ is an equation which can be simplified to the form

$$ax^2 + bx + c = 0,$$

where you wish to solve for x. A typical example is the equation

$$5x^2 - 8x - 4 = 0,$$

which has two solutions: 2 and $-\frac{2}{5}$. (You should check these solutions by substituting in the equation.) A simpler example is $y^2 = 64$, a sentence you can solve in your head. (The solutions are -8 and 8.) Other examples are:

$$\frac{1}{2}y^2 + y = 5y + 6$$

$$\pi r^2 = 14.238$$

The expressions $ax^2 + bx + c$, $\frac{1}{2}y^2 + y$, and πr^2 are $\underline{\text{quadratic}}$ $\underline{\text{expressions}}$. A quadratic expression is a polynomial of degree 2, or an expression which can be simplified to a polynomial of degree 2.

You have seen and will see a variety of applications of quadratic expressions. These applications include:

$\underline{\text{Counting}}$: For n teams to play each other exactly once, $\frac{1}{2}n^2 - \frac{1}{2}n$ games are needed. Similar kinds of formulas exist for other kinds of counting problems, like finding the number of diagonals in a polygon.

Distance: The distance a car, dragster, rocket, or falling object travels depends on speed and acceleration. Acceleration measures how much a rate changes. This "rate of a rate" involves units like feet _per second per second_, or meters per second^2. The second^2 means that the variable for time is often squared in formulas for distance travelled by a moving object.

Statistics: The standard deviation statistic involves quadratic expressions of the form $(x - m)^2$.

Area: You are already familiar with area formulas which involve quadratic expressions. The most famous are the formulas for the area of a square $(A = s^2)$ and for the area of a circle $(A = \pi r^2)$.

The solving of _all_ quadratic sentences (both equations and inequalities) is based upon solving $x^2 = a$. You already know how to do this. We repeat the important definition.

Definition of $\sqrt{}$:

$$\boxed{\text{If } x^2 = a, \text{ then } x = \sqrt{a} \text{ or } x = -\sqrt{a}.}$$

Examples:

1. Solve $\quad y^2 = 4900$

   Solution: $\quad y = \sqrt{4900} \quad$ or $\quad y = -\sqrt{4900}.$

   So $y = 70 \qquad$ or $\quad y = -70.$

   There are usually two solutions.

2. Solve $\quad 3x^2 = 10$

    Solution: $\quad M_{\frac{1}{3}}: \qquad\qquad x^2 = \dfrac{10}{3}$

        Def. of $\sqrt{\ }: \quad x = \sqrt{\dfrac{10}{3}} \quad$ or $\quad x = -\sqrt{\dfrac{10}{3}}$

        The solution set is $\left\{ \sqrt{\dfrac{10}{3}}, -\sqrt{\dfrac{10}{3}} \right\}$.

3. On a Chef Boyardee Cheese Pizza box, the directions read:
    "Spread dough to edges of pizza pan or into a 10" by 14" rectangle
    on cookie sheet."

    Suppose, with this dough, you wanted to make a circular pizza
    with the same thickness. How big a circular pizza could you
    make?

    Solution: Let r be the radius of the circular pizza.

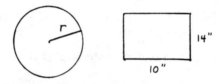

        The area of the circle must equal the area of the
        rectangle.

    In symbols: $\qquad\qquad \pi r^2 = 10 \cdot 14$

    $M_{\frac{1}{\pi}}: \qquad\qquad\qquad\quad r^2 = \dfrac{140}{\pi}$

                              $r^2 \approx \dfrac{140}{3.14} \approx 44.6$

    Def. of $\sqrt{\ }: \qquad\qquad r \approx 6.6$ or $r \approx -6.6$

So the radius is about 6.6". Pizzas are measured by diameter.
So a 10" by 14" rectangular pizza is about the same as a 13"
circular pizza.

<u>Questions covering the reading</u>

1-3. Give an example of a:

1. quadratic expression.   2. quadratic inequality.

3. quadratic equation.

4-9. Find all solutions to each equation.

4. $x^2 = 9$ 5. $1 = y^2$ 6. $z^2 = 80$

7. $2s^2 = 32$ 8. $5b^2 = 0$ 9. $\frac{2}{3}c^2 = 12$

C   10-15. Approximate all solutions to the nearest tenth.

10. $\pi r^2 = 6.28$ 11. $\pi r^2 = 10$ 12. $9 = 11x^2$

13. $v^2 = 11$ 14. $u^2 = \frac{9}{4}$ 15. $2w^2 = 0.75$

16. Pick the two numbers which are solutions to $2x^2 - 3x + 1 = 0$.

     (a) 1   (b) -1   (c) .5   (d) -.5

17. Pick the two numbers which are solutions to $6 = 6m^2 - 5m$.

     (a) 3/2   (b) -3/2   (c) 2/3   (d) -2/3

18-21. Use the expression on p. 652 to calculate the number of games needed for n teams to play each other if n =

18. 6   19. 8   20. 11   21. 24

22. Name four types of applications of quadratic expressions.

23. (a) What equation should you solve to find out the radius of a circular pizza which would have the same amount of pizza as one for a 10'' by 20'' rectangular cookie sheet?

     (b) Solve the equation.

24. Repeat Question 23 for a 25 cm by 40 cm rectangular cookie sheet.

655

25-28. <u>Review</u>. Multiply to show that each expression is a quadratic expression.

25. $(x - 3)(2 - x)$

26. $(2y + 3)^2$

27. $(4 - s)^2$

28. $(z - 5)(2z + 10)$

29-32. <u>Review</u>. Simplify.

29. $4x^2 - 3x + 2x + 8x^2 - 15x + 5 - 3x + 12$

30. $y - y^2 - y^2 - y$

31. $2z^2 - z^2 + 2z - 4z - 2z - z - 2$

32. $412 + 80m - 312m^2 - 150m + 15m^2$

33. What is acceleration?

Questions testing understanding of the reading

1-4. The number of diagonals of a regular polygon with n sides is $\frac{n^2 - 3n}{2}$. Calculate this number for a polygon with

1.   3 sides   2.   4 sides   3.   7 sides   4.   50 sides

5-9. In a vacuum chamber on Earth, an object will drop approximately $4.9t^2$ meters in t seconds. (9.8 meters per second is acceleration due to gravity; 4.9 is half this.)

5.   How far would an object drop in 1 second?

6.   How far would an object drop in 2 seconds?

7.   How far would an object drop in 3 seconds?

C   8.   The Sears Tower in Chicago is (1976) the tallest building in the world and is about 442 meters high. Suppose a vacuum was created in the longest elevator shaft. Then if a penny was dropped from the top of the shaft, how long would it take the penny to reach ground level? (Answer to the nearest second.)

C   9.   Answer Question 8 for the KTHI-TV transmitting tower between Fargo and Blanchard, North Dakota.   This tower is about 629 meters tall.

10-13.   Solve the given equation.   You may have to use the Addition Property of Equations or simplify.

10. $3x^2 + 5 = 5x^2 - 45$

11. $y^2 + 12 - y^2 - y^2 = 12$

12. $(B + 3)(2B + 4) = (B + 5)^2$

13. $256 - m^2 = 0$

Lesson 2

Solving $a(x - h)^2 = b$

These equations are easy to solve. The examples show how.

1. Solve $(t - 3)^2 = 100$.

   Solution: Taking square roots: $t - 3 = 10$ or $t - 3 = -10$

        A_3: $t = 13$ or $t = -7$

   You should check these answers in the original problem.

2. Solve $2(x + 8)^2 = 10$.

   Solution: $M_{\frac{1}{2}}$: $(x + 8)^2 = 5$

            Taking square roots: $x + 8 = \sqrt{5}$ or $x + 8 = -\sqrt{5}$

            A_{-8}: $x = \sqrt{5} - 8$ or $x = -\sqrt{5} - 8$

   Checking: Substitute each solution for x in the original problem.

   $$2(\sqrt{5} - 8 + 8)^2 \overset{?}{=} 10 \qquad 2(\sqrt{5} - 8 + 8)^2 \overset{?}{=} 10$$

   $$2(\sqrt{5})^2 \overset{?}{=} 10 \qquad\qquad 2(-\sqrt{5})^2 \overset{?}{=} 10$$

   $$10 = 10 \qquad\qquad\qquad 10 = 10$$

Expressions of the form $(x - h)^2$ occur often in equations describing paths or orbits of objects, including parabolas, ellipses, and hyperbolas. These expressions are also found in many statistical applications (for instance, the standard deviation calculation) and in describing just about anything which has to do with distance.

1-12. Solve:

1. $x^2 = 9$

2. $(y - 4)^2 = 9$

3. $4(z - 4)^2 = 9$

4. $z^2 - 10 = 0$

5. $(b - 8)^2 = 0$

6. $2(c - 5)^2 = 18$

7. $(m + 2)^2 = 36$

8. $(n + 5)^2 = 25$

9. $3(p + 4)^2 = \frac{1}{3}$

10. $1 = (2u - 3)^2$

11. $9 = (3v + 8)^2$

12. $(5v - 3)^2 = 6$

Questions extending the reading

1. All points on the circle at
    right have coordinates which
    satisfy

    $(x - 2)^2 + (y - 1)^2 = 145.$

    Carefully draw the circle. Its
    center is $(2, 1)$, radius is $\sqrt{145}$.
    By substitution, find the values
    of y which correspond to each
    value of x and put the points
    on the graph.

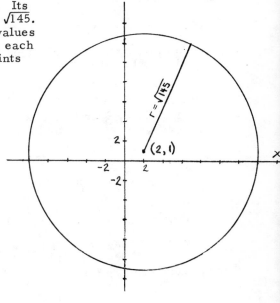

x	y
14	(a)
11	(b)
10	(c)
3	(d)
1	(e)
-6	(f)
-7	(g)
-10	(h)

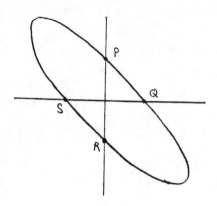

2. The ellipse pictured at left has equation $(x + 2y)^2 + y^2 = 25$.

Find the values of x for each given value of y.

x	y
(a)	0
(b)	3
(c)	-3
(d)	4
(e)	-4
(f)	5

(g) Use these values to give the coordinates of P, Q, R, and S.

Questions for exploration

1-7. These questions explore a process called "completing the square." This process can be used to solve any quadratic equation. Here is an example.

To solve $3x^2 + 5x + 2 = 0$

Step 1: Get all terms in x alone on one side. (Adding -2 to each side will do this.)

$$3x^2 + 5x = -2$$

Step 2: Multiply by 4 times the coefficient of x^2. (In this case, the coefficient is 3, so multiply by 12.)

$$12(3x^2 + 5x) = 12 \cdot -2$$

$$36x^2 + 60x = -24$$

Step 3: Add the square of the original coefficient of x to each side. (The original coefficient is 5, so add 25.)

A_{25}: $36x^2 + 60x + 25 = -24 + 25$

Step 4: The resulting left side will always be the square of a binomial.

$$(6x + 5)^2 = 1$$

Step 5: Solve by the methods of this lesson.

$$6x + 5 = 1 \qquad \text{or} \qquad 6x + 5 = -1$$

$$6x = -4 \qquad \text{or} \qquad 6x = -6$$

$$x = -\frac{2}{3} \qquad \text{or} \qquad x = -1$$

Step 6: Check the solutions you get in the original equation.

1. Complete Step 6 in the example.

2. "Complete the square" to solve $\quad 7x^2 - 8x + 1 = 0$

3. "Complete the square" to solve $\quad 3y^2 + 8y = 28$

4. "Complete the square" to solve $\quad v^2 + 3v + 1 = 0$

5. "Complete the square" to solve $\quad m^2 - 8m + 13 = 0$

6. "Complete the square" to solve $\quad n^2 - 3n + 2 = 0$

7. "Complete the square" to solve $\quad 10x^2 + 3x = 7$

Lesson 3

The Quadratic Formula

If you receive $60 a year on your birthday, then in 3 birthdays you will have $180. You would have more if you put your money in a bank or get interest in some other way. What scale factor will enable you to have $200 by the 3rd birthday?

> Partial
> solution: Let x be the yearly scale factor.
> On the 1st birthday you would have $60.
> On the second birthday you would have $60x + 60$ dollars.
> On the third birthday you have $60x^2 + 60x + 60$ dollars.
> We wish to know the value of x for which
>
> $$60x^2 + 60x + 60 = 200.$$

This problem leads to the solving of a quadratic equation of the form $ax^2 + bx + c = 0$. This would be difficult but there is a formula which gives the solutions to this equation. After the Pythagorean Theorem, this formula is probably the most famous theorem in mathematics. It _must_ be memorized.

Quadratic Formula
    Theorem:

$$\text{If } ax^2 + bx + c = 0, \text{ then}$$

$$x = \frac{-b + \sqrt{b^2 - 4ac}}{2a} \quad \text{or} \quad x = \frac{-b - \sqrt{b^2 - 4ac}}{2a}$$

To apply this formula, notice that a is the coefficient of x^2 and b is the coefficient of x. The constant term is c and 0 must be on one side of the equation.

662

Examples: Applying the Quadratic Formula

1. Solve $3x^2 + 5x + 2 = 0$.

Solution: To apply the formula, notice that a = 3, b = 5, and c = 2.

$$\text{So} \quad x = \frac{-5 + \sqrt{5^2 - 4\cdot3\cdot2}}{2\cdot3} \quad \text{or} \quad x = \frac{-5 - \sqrt{5^2 - 4\cdot3\cdot2}}{2\cdot3}$$

$$= \frac{-5 + \sqrt{25 - 24}}{6} \quad \text{or} \quad = \frac{-5 - \sqrt{25 - 24}}{6}$$

$$= \frac{-5 + 1}{6} \quad \text{or} \quad = \frac{-5 - 1}{6}$$

$$= \frac{-2}{3} \quad \text{or} \quad = -1$$

The solution set is $\left\{\frac{-2}{3}, -1\right\}$. You should check the solutions.

There was a lot of writing in Example 1. The amount of writing can be cut almost in half by using the shorthand symbol $\pm$, read "plus or minus."

$$x \pm y\sqrt{z} \quad \text{is shorthand for} \quad x + y\sqrt{z} \text{ or } x - y\sqrt{z}$$

With this shorthand, the quadratic formula becomes

$$x = \frac{-b \pm \sqrt{b^2 - 4ac}}{2a}$$

2. Solve $5x^2 - 10x - 15 = 0$.

Solution: Here a = 5, b = -10, and c = -15.

$$\text{So} \quad x = \frac{10 \pm \sqrt{100 - 4\cdot5\cdot-15}}{10}$$

$$= \frac{10 \pm \sqrt{400}}{10} = \frac{10 \pm 20}{10}$$

Now translate the shorthand.

$$x = \frac{10 + 20}{10} \qquad \text{or} \qquad x = \frac{10 - 20}{10}$$

$$x = 3 \qquad \text{or} \qquad x = -1$$

Working with an equation before applying the quadratic formula will occasionally help. In Example 2, the numbers could have been made smaller by multiplying both sides of the original equation by $\frac{1}{5}$. In the next example, multiplying both sides by 2 gets rid of fractions.

3. Solve $\frac{1}{2}m^2 - 8m = 5$.

To apply the quadratic formula, 0 must be on one side.

A_{-5}: $\qquad \frac{1}{2}m^2 - 8m - 5 = 0$

M_2: $\qquad m^2 - 16m - 10 = 0$

Now the formula can be easily applied: $a = 1$, $b = -16$, $c = -10$

$$m = \frac{16 \pm \sqrt{(-16)^2 - 4 \cdot 1 \cdot -16}}{2 \cdot 1}$$

$$= \frac{16 \pm \sqrt{256 + 40}}{2}$$

$$= \frac{16 \pm \sqrt{296}}{2}$$

Notice that $\sqrt{296}$ can be simplified. $\sqrt{296} = \sqrt{4} \cdot \sqrt{74} = 2\sqrt{74}$.

So $\qquad m = \frac{16 \pm 2\sqrt{74}}{2}$

$$= 8 \pm \sqrt{74}$$

These are the exact solutions. Decimal approximations can be easily found by using a calculator or by using tables. $\sqrt{74} \approx 8.6$

So $m \approx 8 + 8.6 = 16.6$ $\qquad$ or $\qquad m \approx 8 - 8.6 = -.6$

664

4. Solve the birthday problem on page 662.

$$60x^2 + 60x + 60 = 200$$

A_{-200}: $\quad 60x^2 + 60x - 140 = 0$

$M_{\frac{1}{20}}$: $\qquad 3x^2 + 3x - 7 = 0$

Apply the formula: $\qquad x = \dfrac{-3 \pm \sqrt{9 - 4 \cdot 3 \cdot -7}}{6} = \dfrac{-3 \pm \sqrt{93}}{6}$

So $x = \dfrac{-3 - \sqrt{93}}{6}$, negative solution which does not apply to this problem, or $x = \dfrac{-3 + \sqrt{93}}{6} \approx \dfrac{-3 + 9.464}{6} = \dfrac{6.464}{6} \approx 1.107$

A scale factor of 1.107 is an interest rate of 10.7%, a high rate.

Questions covering the reading

1-8. Each sentence is equivalent to a quadratic equation of the form $ax^2 + bx + c = 0$. Give the values of a, b, and c.

1. $3x^2 + 4x + 1 = 0$

2. $12y + 7y^2 + 5 = 0$

3. $-2m^2 + 8 = 0$

4. $12y^2 - 7y + 1 = 0$

5. $4z^2 - 12z + 9 = 0$

6. $20w^2 - 6w = 2$

7. $1 - x = x^2$

8. $y^2 + 2y + 2 = 5$

9. Name the two solutions to the sentence $ax^2 + bx + c = 0$.

10. What does "$\pm$" mean? How does this symbol shorten the Quadratic Formula?

11-18. Solve the equations of Questions 1-8.

19-22. Suppose that you deposit \$50 in a bank every year on your birthday. The bank will, in effect, multiply this money each year by a scale factor of x.

19. How much will you have by the 3rd birthday?

20. What value of x will enable you to have \$150 by the 3rd birthday?

C 21. What value of x will enable you to have \$160 by the 3rd birthday? (Answer to the nearest hundredth.)

C 22. What value of x will enable you to have \$165 by the 3rd birthday? (Answer to the nearest hundredth.)

23-34. Find all solutions to each equation.

23. $18t^2 - 39t + 20 = 0$

24. $0 = 180t^2 - 390t + 200$

25. $x^2 + 9x + 20 = 0$

26. $10y - 5 = y^2$

27. $8z^2 - 8z + 2 = 0$

28. $3w^2 + 82w + 5 = 4w^2 + 94w + 6$

29. $A(A + 10) = 39$

30. $4B(B - 4) = 9$

31. $50m^2 + 60m + 10 = 0$

32. $0 = \frac{1}{4}x^2 + \frac{2}{3}x + \frac{1}{6}$

33. $3 = 4 - 5n - 6n^2$

34. $cx^2 + bx + a = 0$ (Solve for x.)

Questions testing understanding of the reading

1-2. The following problems are taken almost verbatim from Introduction to Mathematics for Life Scientists, by E. Batschelet, p. 89.

1. In 1964, S. Wright (Statistics and mathematics in biology, Chapter 2, p. 27) in an analysis of heredity was led to the equation $4x^2 - 2x - 1 = 0$. Find the roots.

2. In 1965, R. A. Fisher (The theory of inbreeding, p. 64), discussing the breeding of animals with long pregnancy and only one offspring at birth, considered the equation $8\lambda^2 - 8\lambda + 1 = 0$, where λ is the Greek letter lambda. Solve this equation.

666

3-4. (These problems are from <u>Introductory Mathematics for Technicians</u>, by Alvin Auerbach and Vivian Groza, p. 586-595.) If you move a thermometer from a room whose temperature is 70°F to an outside location whose temperature is 45°F, then it takes time for the thermometer's reading to change from 70°F to 45°F. If t is the time in minutes, then the thermometer reading T is approximated by

$$T - 45 = 25(1 - .04t)^2$$

3. How many minutes will it take the thermometer to get down to 46°F?

C 4. How long will it take the thermometer to get down to 45.5°F?

5. A bullet is fired upward from ground level with an initial speed (called <u>muzzle velocity</u>) of 670 meters per second. This speed, gravity, and air resistance affect how long the bullet will be in the air (and how high it will go).

    (a) If there were no gravity or air resistance, how high would the bullet be after t seconds?

    (b) On Earth, gravity lowers the bullet about $5t^2$ meters in t seconds. So, allowing for gravity, how high will the bullet be after t seconds?

C   (c) If there were no air resistance, when would the bullet be a kilometer above the ground?

C   (d) If there were no air resistance, when would the bullet hit the ground? (Air resistance has a great effect on the time the bullet is in the air, as this question shows.)

<u>Questions for discussion</u>

1. Steps (a) to (g) show how the quadratic formula can be worked out. Tell what was done to get each step.

$$ax^2 + bx + c = 0$$

(a) $4a^2x^2 + 4abx + 4ac = 0$

(b) $4a^2x^2 + 4abx \quad\quad = -4ac$

(c) $4a^2x^2 + 4abx + b^2 = b^2 - 4ac$

(d) $(2ax + b)^2 = b^2 - 4ac$

667

(e) $\quad 2ax + b = \sqrt{b^2 - 4ac}$ $\quad$ or $\quad 2ax + b = -\sqrt{b^2 - 4ac}$

(f) $\quad 2ax = -b + \sqrt{b^2 - 4ac}$ $\quad$ or $\quad 2ax = -b - \sqrt{b^2 - 4ac}$

(g) $\quad x = \dfrac{-b + \sqrt{b^2 - 4ac}}{2a}$ $\quad$ or $\quad x = \dfrac{-b - \sqrt{b^2 - 4ac}}{2a}$

2. There is a relationship between steps (a) and (g) of Question 1 and the process of "completing the square" found in Questions for exploration, p. 660. What is this relationship?

3. In Greek architecture and many paintings, a rectangle often used to outline buildings or people or objects is similar to the ones drawn below. Rectangles of this shape are called <u>golden rectangles</u>.

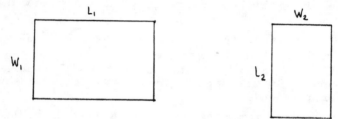

In golden rectangles, the length and width satisfy the equation

$$\frac{L + W}{L} = \frac{L}{W}$$

That is, the length is the geometric mean of the width and the sum of the length and width.

(a) Why is the formula equivalent to

$$1 + \frac{W}{L} = \frac{L}{W} \ ?$$

(b) Let $x = \frac{L}{W}$, the ratio of length to width. Then, substituting:

$$1 + \frac{1}{x} = x$$

Where does the $\frac{1}{x}$ come from?

(c) Multiplying both sides by x:

$$x + 1 = x^2$$

Show that $x \approx 1.618$.

(d) If a building is 100 meters wide and its front is shaped like a golden rectangle, how high is the building?

(e) Many windows are shaped like golden rectangles. Measure a window where you live. How close is $\frac{L}{W}$ to 1.618?

(f) Look for other rectangular-shaped objects where you live (TV screens, oven doors, etc.). Can you find a rectangle with $\frac{L}{W}$ close to 1.618?

Lesson 4

The Zero Product Theorem

When some numbers add to zero, you cannot be sure what they are. Suppose

$$w + x + y + z = 0$$

Then it is possible that $w = 419$, $x = -1200$, $y = -2$, and $z = 783$.
You could not predict any of these values in advance.

But if numbers <u>multiply</u> to zero,

$$w \cdot x \cdot y \cdot z = 0$$

then one thing can be said about them. At least one of the numbers itself must be zero.

<table>
<tr><td>

<u>Zero Product
Theorem</u>:

</td><td>

If a product of numbers is zero, then at least one of the factors is zero. Symbolically:
If $ab = 0$, then $a = 0$ or $b = 0$.

</td></tr>
</table>

Here are two ways to show that the Zero Product Theorem is true.

(1) Suppose a and b stand for the length and width of a rectangle. Then ab is the area. The area is 0 only when one of the sides is 0.

(2) Think of solving the equation ab = 0 for b. Now either a = 0 or a ≠ 0. If a is not equal to zero, then you can multiply both sides by $\frac{1}{a}$ and find that b = 0. So either a = 0 or b = 0.

The Zero Product Theorem can be applied to solve equations when one side is 0 and the last operation on the other side is multiplication.

Examples: 1. Solve: (x - 2)(3x + 12) = 0

Solution: The right side is 0. The last operation on the left side is multiplication. (It's the only operation left after you have worked in the parentheses.) So the Zero Product Theorem can be applied. Using that theorem,

either   x - 2 = 0   or   3x + 12 = 0

So   x = 2   or   3x = -12

So   x = 2   or   x = -4

The solution set is $\{2, -4\}$.

2. When can the area of a circle be zero? (The obvious answer to this question can be verified using equations.)

Answer: The area of a circle with radius r is πr^2.

We want to know when $\pi r^2 = 0$

By the Zero Product Theorem,

either π = 0   or r^2 = 0

       π = 0   or   r = 0

671

But $\pi = 3.14...$ So $\pi \neq 0$. So it must be that $r = 0$. This shows that a circle has zero area only when its radius is 0. (Sometimes a person's intuition about area is not correct. Equation-solving guarantees that no surprise answer exists.)

Questions covering the reading

1. According to the Zero Product Theorem, if $xy = 0$, then _____ or _____.

2. If $0 = (x - 2)(x + 6)$, then _____ or _____.

3. How can a rectangle be used to verify the Zero Product Theorem?

4. Given $a \neq 0$, solve for x: $ax = 0$.

5-10. Find all solutions to each equation.

5. $(x - 5)(x - 3) = 0$ 6. $(y + 7)(y - 2) = 0$

7. $(3a - 6)(4a - 1) = 0$ 8. $(2B + 1)(B + 14) = 0$

9. $n(n + 3) = 0$ 10. $x(2x - 3) = 0$

11-12. Why can't the Zero Product Theorem be used to help solve the given equation?

11. $ab = 100$ 12. $(x - 5) + (2x - 3) = 0$

13. To find out when the area of a circle is zero, what equation might you solve?

14. When does a circle have zero area?

672

Questions testing understanding of the reading

1-8. Find all solutions to each equation.

1. $(3A - 5)(2A + 3) = 0$

2. $(3 - c)(2 + 5c) = 0$

3. $(m - 2)(m + 3)(m - 4) = 0$

4. $0 = (t - 2)(5 + t)(3t - 1)$

5. $x(7x + 1) = 0$

6. $n^2(2n - 6) = 0$

7. $100t(t + 1) = 0$

8. $75m^3(5m - 9) = 0$

9-14. Give an equation which has all of the given solutions.

9. $x = 2$ or $x = 6$

10. $y = 9$ or $y = -\frac{1}{2}$

11. $z = -\frac{3}{4}$ or $z = \pi$

12. $c = 6$ or $c = -6$

13. $a = 0$ or $a = 1$

14. $n = -11$ or $n = 3$ or $n = 0$

15-16. When does the given expression not stand for a number?

15. $\dfrac{(x - 1)(x - 2)}{(x - 3)(x - 4)}$

16. $\dfrac{y(y + 6)}{(y + 2)(2y - 3)}$

Questions for exploration

1. The area A of a trapezoid (drawn below) is given by the formula

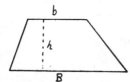

$$A = \tfrac{1}{2}h(b + B)$$

where b and B are the small and large bases and h is the height.

(a) Show that the area can be 0 only if the height is 0 or <u>both</u> bases are 0.

(b) The answer to (a) leaves open the possibility that one base could be 0 and yet there would still be a positive area. Draw such a trapezoid.

2. Suppose you want to say "x = 3 or y = 5" but you want to use only one = sign. You could write (x - 3)(y - 5) = 0.

    (a) Using only one equal sign, write: x = 6 or y = 2 or z = -5.

    (b) Using only one equal sign, write: x = y or x = z.

    (c) Make up your own example like part (a) or part (b).

Lesson 5

Factoring Polynomials

If you begin with <u>3 times 2</u> and end with 6, the process is called multiplication. The reverse process, beginning with 6 and ending with <u>3 times 2</u>, is called <u>factoring</u>. The same words are used with polynomials. If 3 is multiplied by (n - 7), the product is <u>3n - 21</u>. So:

Example 1:
$$3n - 21 = 3(n - 7)$$
↑   ↑

      original   factored
   polynomial   form

All factoring is based on the distributive property. Here is an example where each side has a "common sense" meaning.

Example 2:
$$500x^3 + 500x^2 + 500x + 500 = 500(x^3 + x^2 + x + 1)$$

Suppose you paid $500 into a fund for three years and each year the fund multiplied the amount by x. Then with the fourth year's payment you would have the <u>left</u> side expression

$$500x^3 + 500x^2 + 500x + 500.$$

If 500 people each paid $1 a year into the same fund, the same total would result. This total is given by the <u>right</u> side.

$$500(x^3 + x^2 + x + 1)$$

Whenever each term of a polynomial has the same factor, the entire polynomial can be factored.

Examples: Factoring Polynomials

3. $2x^2 + 4x + 20 = 2(x^2 + 2x + 10)$

4. $14y - 35y^2 + 77 = 7(2y - 5y^2 + 11)$

5. The common factor can contain variables.

$9a - 15ab = 3a(3 - 5b)$

6. $3x^5y + x^3y = x^3y(3x^2 + 1)$

Some polynomials can be factored even though there is not a common factor in each term. In Chapter 12, the FOIL Theorem showed that

$$(x - y)(x + y) = x^2 - y^2$$

Reversing the equality shows how to factor <u>the difference of two squares</u>:

$$x^2 - y^2 = (x + y)(x - y)$$

Examples: Factoring the Difference of Two Squares:

7. To factor $9 - y^2$, notice that each term is a perfect square.

$$9 - y^2 = 3^2 - y^2$$
$$= (3 + y)(3 - y)$$

8. $64x^2 - 25 = (8x)^2 - 5^2$
$$= (8x + 5)(8x - 5)$$

9. $4z^2 - 1 = (2z)^2 - 1^2$
$$= (2z + 1)(2z - 1)$$

The two types of factoring can be combined in one situation. The next example is famous.

10. The area of a ring (shaded at right) is given by the formula
$$A = \pi R^2 - \pi r^2$$

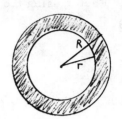

The number π is a common factor.
$$A = \pi(R^2 - r^2)$$

Factoring the difference of two squares,
$$A = \pi(R + r)(R - r)$$

(You first saw this formula in Chapter 2.)

Questions covering the reading

1-12. Factor.

1. $9m + 9n$

2. $40x^2 + 40x + 40$

3. $12 - 36y$

4. $120 - 360z^2$

5. $30x + 60t$

6. $25a - 5b$

7. $6m + 6m^2 + 6m^3$

8. $x + xy + 3xz$

9. $2n^2 - 20mn$

10. $2v^2 - v$

11. $w + w^4$

12. $3m^2n^3 + mn^2 - mn^4$

13. When factored, $x^2 - y^2 =$ _____.

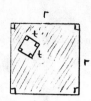

14. <u>Review</u>. The area of the shaded region at left is _____; in factored form this area is

_____.

15-24. Factor using the difference of two squares.

15. $a^2 - 81$ 16. $b^2 - 16$

17. $c^2 - 4d^2$ 18. $9e^2 - 1$

19. $900f^2 - 625$ 20. $16 - 25g^2$

21. $u^2 - 49$ 22. $v^2 - 36w^2$

23. $\frac{1}{4}x^2 - \frac{4}{9}$ 24. $121 - 169y^2$

25. In two different ways, write the formula for the area of a ring.

26. Calculate the area of a 5 foot wide circular walk which is built around a circular garden of radius 47.5 feet. (Hint: It's easy to calculate in factored form.)

27-30. Factor as much as possible.

27. $200x^2 - 8$ 28. $40 - 160y^2$

29. $75a^2 - 75b^2$ 30. $3m^2 - 147n^2$

<u>Questions testing understanding of the reading</u>

1. The number of diagonals in an n-sided polygon is

$$\frac{n^2 - 3n}{2}$$

Factor the numerator of this fraction.

2. A league has t teams. If each team is to play each other home
    and away,

$$t^2 - t$$

    games are needed. Factor this expression.

3-4. The amount of concrete needed
for the cylindrical sewer pipe pictured
at right is

$$\pi R^2 L - \pi r^2 L,$$

where R is the outer radius, r is the
inner radius, and L is the length.

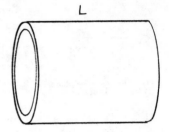

L

3. Factor this expression.

C 4. Find the amount of concrete needed for a sewer pipe 2500 meters
    long with outer radius of 2 meters and inner radius of 1.9 meters.
    Use the approximation 3.14 for π.

5-8. When does the expression not stand for a number?

5. $\dfrac{1}{m^2 - 4}$

6. $\dfrac{3}{9x^2 - 4}$

7. $\dfrac{2(y + 6)}{(y + 1)(y^2 - 9)}$

8. $\dfrac{t^2 - 16}{t^2 - 25}$

9-14. Factor the left side of each equation. Then solve the equation
which remains either using the Zero Product Theorem or the Quadratic
Formula.

9. $500x^2 + 1000x - 1500 = 0$

10. $-60 + 12z + 12z^2 = 0$

11. $9y^2 - 16 = 0$

12. $100m^2 - 4 = 0$

13. $60 - 60n^2 = 0$

14. $52 - 13t^2 = 0$

15-20. (a) Factor the numerator and denominator. (b) Then apply
the Fraction Simplification Property first discussed on p. 239 to
simplify the fraction.

15. $\dfrac{7x + 14}{x^2 - 4}$

16. $\dfrac{y^2 - 9}{2y + 6}$

17. $\dfrac{5 - 3m}{9m^2 - 25}$

18. $\dfrac{4z + 1}{16z^2 - 1}$

19. $\dfrac{18 - 2y^2}{9 + 3y}$

20. $\dfrac{200n^2 - 50m^2}{40m + 80n}$

21. P and Q are joined by an arc
with center O and by the segment $\overline{PQ}$. The area of the
region between the arc and the
segment is

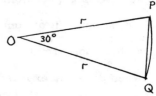

$$\frac{\pi r^2}{12} - \frac{r^2}{4}.$$

(a) Factor this expression.

(b) Use the factored form to find the area when r = 70. Use the
approximation 22/7 for π.

Lesson 6

Factoring Quadratic Expressions

Many algebra courses spend many weeks on factoring. This is because factoring is a process by which many equations can be solved. But better ways are known and computers use other methods for sentence-solving. So in this book, only a few lessons are devoted to factoring.

Still, factoring is helpful in

(a) evaluating some formulas (for examples, see Question 4, p. 679, or Question 21, p. 680).

(b) simplifying some expressions (for examples, see Questions 15-20, p. 680).

and (c) learning about polynomials. That is what this lesson is for.

Given $(x - 4)(x - 1)$

you can apply the FOIL Theorem to get

$$x^2 - 5x + 4$$

But suppose you were given $\underline{x^2 - 5x + 4}$. It is possible to figure out that the factors are $\underline{x - 4}$ and $\underline{x - 1}$. The quadratic formula helps in doing this.

Let $x^2 - 5x + 4 = 0$

Then $x = \dfrac{5 \pm \sqrt{25 - 4 \cdot 4 \cdot 1}}{2}$

So (after some computation)

$$x = 4 \quad \text{or} \quad x = 1$$

So $\qquad x - 4 = 0 \quad \text{or} \quad x - 1 = 0$

This shows that $(x - 4)(x - 1) = 0$ and suggests the factoring of $x^2 - 5x + 4$.

$$x^2 - 5x + 4 = (x - 4)(x - 1).$$

Here is the general pattern.

Examples: Factoring Quadratics

1. Suppose you want to factor $4x^2 - 7x + 3$. First solve $4x^2 - 7x + 3 = 0$. By the Quadratic Formula,

$$x = 1 \quad \text{or} \quad x = \frac{3}{4}$$

So $\qquad x - 1 = 0 \quad \text{or} \quad x - \frac{3}{4} = 0$

That is, $\quad (x - 1)(x - \frac{3}{4}) = 0$

We need $4x^2$, so multiply by 4.

$$4(x - 1)(x - \frac{3}{4}) = 0$$

This means $\quad 4x^2 - 7x + 3 = 4(x - 1)(x - \frac{3}{4})$

To get rid of fractions, multiply the 4 and the $x - \frac{3}{4}$.

$$= (x - 1)(4x - 3)$$

2. To factor $8x^2 + 10x + 3$, first solve $8x^2 + 10x + 3 = 0$.

By the formula, $x = -\dfrac{3}{4}$ or $x = -\dfrac{1}{2}$

So $x + \dfrac{3}{4} = 0$ or $x + \dfrac{1}{2} = 0$

That is, $\left(x + \dfrac{3}{4}\right)\left(x + \dfrac{1}{2}\right) = 0$

So $8\left(x + \dfrac{3}{4}\right)\left(x + \dfrac{1}{2}\right) = 0$

So $8x^2 + 10x + 3 = 8\left(x + \dfrac{3}{4}\right)\left(x + \dfrac{1}{2}\right)$

$$= 4\left(x + \dfrac{3}{4}\right) \cdot 2\left(x + \dfrac{1}{2}\right)$$

$$= (4x + 3)(2x + 1)$$

Examples 1 and 2 show how factors of a quadratic expression are related to the solutions to the corresponding quadratic equation.

Solution	Factor
1	x - 1
$\dfrac{3}{4}$	4x - 3
$-\dfrac{3}{4}$	4x + 3
$-\dfrac{1}{2}$	2x + 1

There is a simple relationship.

<u>Factor Theorem for Quadratic Expressions:</u> | If $\dfrac{n}{d}$ is a solution to $ax^2 + bx + c = 0$, then $(dx - n)$ is a factor of $ax^2 + bx + c$.

Suppose the solution $\frac{n}{d}$ is a rational number. Then n and d

may be integers. So dx - n will be simple. A quadratic equation

with rational solutions means nice factors for the expression. But

many quadratics have irrational solutions. Then the factors are

quite complicated.

Example 3. The solutions to $x^2 - 4x + 2 = 0$ are $2 - \sqrt{2}$ and $2 + \sqrt{2}$.

So $x - (2 - \sqrt{2}) = 0$ or $x - (2 + \sqrt{2}) = 0$

That is,

$(x - (2 - \sqrt{2}))(x - (2 + \sqrt{2})) = 0$

Thus $x^2 - 4x + 2 = (x - (2 - \sqrt{2}))(x - (2 + \sqrt{2}))$

$= (x - 2 + \sqrt{2})(x - 2 - \sqrt{2})$

There are no simpler factors. The polynomial $x^2 - 4x + 2$

is not factorable using only integers.

Questions covering the reading

1-4. Apply the distributive property.

1. $5(y - \frac{2}{5})$ 2. $4(x + \frac{1}{4})$

3. $10(z + \frac{1}{3})$ 4. $2(m - \frac{17}{2})$

5-10. Fill in the blanks.

5. If $x = 10$, then $x - 10 =$ _____ .

6. If $x = -3$, then $x +$ _____ $= 0$.

7. If $y = 2$ or $y = 4$, then $($ _____ $)($ _____ $) = 0$.

8. If $z = 6$ or $z = -1$, then $($ _____ $)($ _____ $) = 0$.

9. If $w = -2$ or $w = \frac{2}{3}$, then $($ _____ $)($ _____ $) = 0$.

10. If $x = -\frac{1}{4}$ or $x = -6$, then $($ _____ $)($ _____ $) = 0$.

11-18. A quadratic equation and its solutions are given. Use this information to factor the left side of the equation.

11. $x^2 - 5x + 4 = 0$ Solutions: 1 and   4

12. $y^2 + 20y + 99 = 0$ Solutions: -11 and -9

13. $z^2 - 8z - 9 = 0$ Solutions: -1 and   9

14. $w^2 + 3w - 28 = 0$ Solutions: -7 and   4

15. $2x^2 - 3x + 1 = 0$ Solutions: 1 and $\frac{1}{2}$

16. $5y^2 - 7y - 6 = 0$ Solutions: $-\frac{3}{5}$ and   2

17. $4z^2 + 8z + 3 = 0$ Solutions: $-\frac{3}{2}$ and $-\frac{1}{2}$

18. $18w^2 + 3w - 1 = 0$ Solutions: $\frac{1}{6}$ and $-\frac{1}{3}$

19. Name 3 ways in which factoring is helpful.

20. State the Factor Theorem for quadratic expressions.

21-24. Each quadratic equation has no rational solutions. Use this information to factor the left side of the equation.

21. $x^2 - x + 1 = 0$

22. $x^2 - x + 3 = 0$

23. $x^2 - 4x - 2 = 0$

24. $x^2 - 3x - 3 = 0$

25-30. Factor.

25. $x^2 + 15x + 56$

26. $6y^2 - 17y + 12$

27. $z^2 - 2z - 3$

28. $3m^2 + 2m - 1$

29. $20n^2 + 44n - 15$

30. $7x^2 - 91x + 12$

Questions extending the reading

1-6. Tell whether the given expression is factorable using only rational coefficients.

1. $3x^2 + 11x + 6$

2. $3x^2 + 10x + 6$

3. $3x^2 + 9x + 6$

4. $2y^2 - 5y + 3$

5. $2y^2 - 6y + 3$

6. $2y^2 - 7y + 3$

7-14. Simplify the fraction by first (a) factoring the numerator and denominator and then (b) applying the Fraction Simplification Property.

7. $\dfrac{x^2 + 3x + 2}{x^2 - 4}$

8. $\dfrac{9 - y^2}{y^2 + 4y + 3}$

9. $\dfrac{5m^2 - 7m + 2}{3m - 3}$

10. $\dfrac{40 + 100x}{14 + 31x - 10x^2}$

11. $\dfrac{3z^2 - 4z + 1}{3z^2 + 2z - 1}$

12. $\dfrac{2n^2 + n - 1}{4n^2 - 6n + 2}$

13. $\dfrac{x^3 + 2x^2 + x}{x^2 - 1}$

14. $\dfrac{30m^2 - 71m + 42}{25m^2 - 36}$

15. The following problem is quoted from <u>Introductory Mathematics</u>
<u>for Technicians</u>, by Auerbach and Groza, p. 179.

"The bending moment for a certain beam of
length L, uniformly loaded   w lb per unit
length, at a distance x from one end is
given by

$$M = \tfrac{1}{2}wx^2 - \tfrac{5}{8}wLx + \tfrac{1}{8}wL^2$$

Write M in factored form."

Let w = 1 and L = 1.   (a) What does M then equal?

(b) Calculate and factor 8M.

Questions for exploration

1-2. Some very simple polynomials have many factors. Check the
given factoring by letting x = 3 and then letting x = 2.

1. $x^4 - 16 = (x^2 + 4)(x - 2)(x + 2)$

C 2. $x^6 - 1 = (x - 1)(x + 1)(x^2 + x + 1)(x^2 - x + 1)$

Lesson 1: Quadratic Expressions
      2: Solving $a(x - h)^2 = b$
      3: The Quadratic Formula
      4: The Zero Product Theorem
      5: Factoring Polynomials
      6: Factoring Quadratic Expressions

Every quadratic equation can be simplified to the form

$$ax^2 + bx + c = 0$$

The <u>Quadratic Formula</u> gives the solutions to this equation.

$$x = \frac{-b \pm \sqrt{b^2 - 4ac}}{2a}$$

Some quadratic equations are in other forms than the one above and can more easily be solved without using the formula. One such form is

$$a(x - h)^2 = b,$$

a type of sentence which often occurs in graphs and in statistics. Another such form is the <u>factored form</u>

$$(x - a)(x - b) = 0.$$

To solve equations of this form, the Zero Product Theorem can be applied. If the product of two or more numbers is zero, then at least one of them must be zero. So, for the above equation, either $x - a = 0$ or $x - b = 0$. So $x = a$ or $x = b$.

Three types of factoring are discussed in this chapter: (1) factoring a polynomial where each term has the same factor; (2) factoring the difference of two squares using the theorem

$$x^2 - y^2 = (x + y)(x - y),$$

and (3) factoring the quadratic expression $ax^2 + bx + c$ by using the ' Quadratic Formula. Factoring is helpful in evaluating some formulas, simplifying some expressions, and in the study of polynomials.

CHAPTER 16

FUNCTIONS

Lesson 1

The Definition of Function

In earlier lessons, you have seen many types of graphs. Here are four:

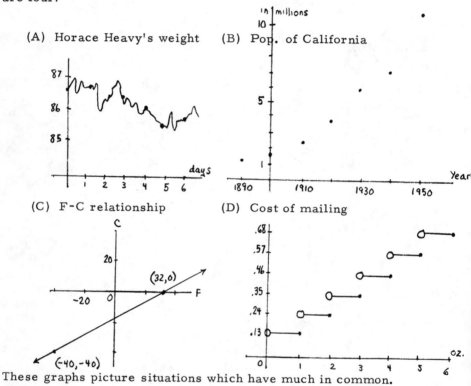

(A) Horace Heavy's weight

(B) Pop. of California

(C) F-C relationship

(D) Cost of mailing

These graphs picture situations which have much in common.

In each situation:

(1) There are two variables. (The first variable is

indicated horizontally; the second vertically.)

(2) Every value of the first variable determines exactly

one value of the second variable.

For example, in graph (D),p. 690, the first variable stands for weight,

the second for cost. The weight determines the cost of mailing. (The

cost of mailing does not determine the weight.)

When a situation has two variables, it is natural to think of

describing the situation using ordered pairs. The general name

given to sets of ordered pairs like those above is <u>function</u>.

<table>
<tr><td>Definition:</td><td>A <u>function</u> is a set of ordered pairs
in which each first coordinate appears
with exactly one second coordinate.</td></tr>
</table>

The set of ordered pairs

graphed at right is <u>not</u> a function.

The first coordinate 4 appears

with two different second coordi-

nates, 2 and -2.

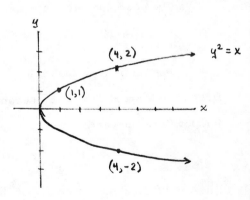

The idea of a function is very simple. If you have two variables so related that the first determines the second, there is a function around. It is the set of ordered pairs of related things.

Because the idea is simple, functions are found in every branch of mathematics and are extremely important in many applications. As a result, there are many ways to describe functions.

Ways of describing functions

1. By a graph. Four functions are pictured on page 690.
2. By a written rule. The mailing cost function (D) can be described as follows: The cost is 13¢ for the first ounce and 11¢ for each additional ounce up to 14 ounces.
3. By listing the pairs. The population of California function (B) contains the pairs (1890, 1213398), (1900, 1485053), etc.
4. By an equation. The Fahrenheit-Celsius conversion function (C) can be described by the equation $C = \frac{5}{9}(F - 32)$. Using the common variables x and y, $y = \frac{5}{9}(x - 32)$.

In descriptions of functions, it is common to let x stand for the first variable and y for the second variable. For example, in the temperature conversion function, let x = F and y = C. Then

$$y = \frac{5}{9}(x - 32)$$

The use of x and y shows that the graph of this function is a line.

Questions covering the reading

1. Define: function.

2. Give an example of a set of ordered pairs which is not a function.

3-7. Tell whether the set of ordered pairs is a function.

3. $\{(2,3), (3,2), (1,4), (4,1)\}$ 4. $\{(-6,10), (-5,10), (-4,10)\}$

5. $\{(.1,6), (.1,7), (.1,8)\}$ 6. $\{(0,3), (3,0), (0,-4)\}$

7. $\{(-1,-2), (-2,-3), (-3,-4), (-4,-5), \ldots\}$

8. Name four ways in which a function can be described.

9-12. Name three ordered pairs (if possible) of these functions graphed on page 690.

9. Function (A): Horace Heavy's weight function (Refer to p. 393)

10. Function (B): Population of California (Refer to p. 534)

11. Function (C): F-C relationship

12. Function (D): Mailing cost function

13. Give a written rule for the mailing cost function (D) of p. 690.

14. Give an equation for the F-C relationship.

15. In graphing it is customary to let _____ stand for the first variable.

16. In graphing it is customary to let _____ stand for the second variable.

17-20. Each equation describes a function. Name four ordered pairs of the function.

17. $y = -4x^2$

18. $2x - y = 6$

19. $y = 4x - 2$

20. $xy = 15$

Questions extending the reading

1. A <u>continuous function</u> is one whose graph is continuous. (See page 393 if you have forgotten what this means.) Which of the functions on page 690 are continuous?

2. A <u>discrete function</u> is one whose graph is made up of unconnected points. Which functions on page 690 are discrete?

3-5. Each sentence could be a written rule for a function. Name three ordered pairs of the function.

3. Every person has an age.

4. These potatoes cost 10¢ a pound.

5. For every cm of thickness of insulation, only $\frac{2}{5}$ of sound gets through.

Lesson 2

Function Notation

You are familiar with the abbreviation P(E). In this abbreviation, the letter E names an event and P(E) names the probability of that event. An event has only one probability, so any set of events and their probabilities is a function. This kind of abbreviation is used for all functions. We can use the shorthand s(x) to stand for the square of x. ("s(x)" is read "s of x.") Then for each x, there is s(x). This leads to a squaring function.

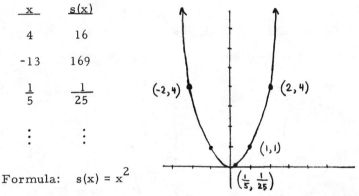

x	s(x)
4	16
-13	169
$\frac{1}{5}$	$\frac{1}{25}$
:	:

Formula: $s(x) = x^2$

Ordered pairs of the squaring function: (4, 16), (-13, 169)

$(-2, 4)$, $(\frac{1}{5}, \frac{1}{25})$, ...

In the abbreviation s(x), the parentheses again do not mean multiplication. s(x) stands for the second variable in the function with first variable x. Each ordered pair of the squaring function

695

has the form $(x, s(x))$. If $x = 3$, $s(x) = x^2 = 3^2 = 9$.

Examples: Function notation

1. Consider a function with equation $f(n) = n^2 + n$. Points of the function are found by substitution.

<u>$f(n) = n^2 + n$</u>   <u>Elements of the function</u>

$f(0) = 0^2 + 0 \quad = 0$ $(0, 0)$

$f(6) = 6^2 + 6 \quad = 42$ $(6, 42)$

$f(-6) = (-6)^2 + -6 = 30$ $(-6, 30)$

$f(\pi) = \pi^2 + \pi$ $(\pi, \pi^2 + \pi)$

$f(\frac{2}{3}) = (\frac{2}{3})^2 + \frac{2}{3} \quad = \frac{10}{9}$ $(\frac{2}{3}, \frac{10}{9})$

2. The equation <u>$t(x) = -5x + 40$</u> describes the linear function with slope -5, y-intercept 40.

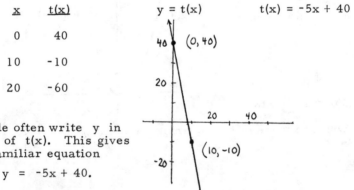

x	t(x)
0	40
10	-10
20	-60

$y = t(x)$ $t(x) = -5x + 40$

People often write y in place of t(x). This gives the familiar equation

$y = -5x + 40$.

Sometimes a single letter is used to name the function. The rule in Example 2 uses t(x). So the function is called t. The rule in Example 1 uses f(x). So the function is called f. The most common letter used is f.

696

1-5. Consider the function g with rule g(x) = 5x + 3. Fill in the blank.

1. g(3) = _____   2. g(10) = _____   3. g(-10) = _____

4. When x = 10, g(x) = _____.

5. When x is 1.2, g(x) is _____.

6-10. Name 4 ordered pairs in the function with the given equation.

6. $f(t) = \frac{5}{9}(t - 32)$ 7. $L(x) = 2 + 4x$

8. $p(x) = 10x - 4x^2$ 9. $g(a) = a^2 + a^3$

10. $f(m) = m(m - 2)$

11-15. Repeat Questions 1-5 for the function with equation

$$g(x) = |x - 5|.$$

16-17. Suppose h(a) = a + 3 is an equation for a function.

16. Name three ordered pairs of this function.

17. The graph of this function is a _____.

Questions testing understanding of the reading

1-3. Refer to the functions of Questions 6-10, above.

1. Which two functions might be called linear functions?

2. Which function might be called a temperature function? Why?

3. Which function would be called a polynomial function of degree 3?

4-6. Calculate f(1), f(2), f(3), f(4), f(5), and f(6) when

4. $f(x) = 100x - 2$ 5. $f(x) = -(x - 3)^2$

6. $f(x) = |x - 3|$

7-14. Let S be the set of all people. For each element x of S
(when appropriate), let f(x) = the father of x, m(x) = the mother of x,
h(x) = the husband of x, w(x) = the wife of x, s(x) = the eldest sister
of x, and b(x) = the youngest brother of x. What name should be
given to:

7. f(f(x))   8. f(m(x))   9. m(f(x))   10. m(m(x))

11. s(f(x))   12. b(f(x))   13. m(w(x))   14. f(h(x))

15. If $s(n) = n^2$, simplify s(s(s(n))).

16-19. Let f(t) = 9 + 5t.

16. Solve for t: f(t) = 0.   17. Solve for t: f(t) = 60.

18. f(a) = _____.   19. f(a+2) = _____.

20-23. Repeat Questions 16-19 if $f(t) = 3t^2 - 4$.

Lesson 3

Linear Functions

Functions are found in every part of mathematics. In the next five lessons, many things you have learned are reviewed using the language and notation of functions.

Idea	Type of Function
lines	linear functions (this lesson)
absolute value	absolute value functions (Lesson 4)
quadratic expressions	quadratic functions (Lesson 5)
probability	probability functions (Lesson 6)
frequencies	frequency distributions (Lesson 7)

The most basic of all functions are the linear functions. A linear function is a function whose graph is all or part of a line.

Recall that every non-vertical line has an equation of the form

$$y = mx + b$$

where m is the slope and b is the y-intercept of the line. In function language, you could write

$$f(x) = mx + b$$

Every linear function has an equation of this type.

For example, $f(x) = -2x + 3$
is an equation for the linear
function f with slope -2 and
y-intercept 3.

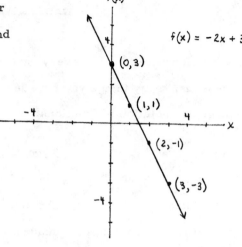

The slope of the line through (x_1, y_1) and (x_2, y_2) is

$$\frac{y_2 - y_1}{x_2 - x_1}$$

When the line is the graph of a function f, then $y_1 = f(x_1)$ and
$y_2 = f(x_2)$. Substituting, the slope formula becomes

$$\frac{f(x_2) - f(x_1)}{x_2 - x_1}$$

For example, when $f(x) = -2x + 3$, let $x_1 = 5$ and $x_2 = 10$.

Then $\qquad f(x_1) = -2 \cdot 5 + 3 \ = -7$

$\qquad\qquad f(x_2) = -2 \cdot 10 + 3 = -17$

The slope is $\dfrac{-17 - -7}{10 - 5}$ or $\dfrac{-10}{5}$ or -2, which could have been predicted
from the equation for f.

Every function is a set of ordered pairs. The set of first coordinates of these ordered pairs is called the <u>domain</u> of the function. Here are two functions with the same function rule. But the domains are different.

Function f: f(x) = 2x

f(x) is the price of two basketballs if one costs x dollars.

Graph of f:

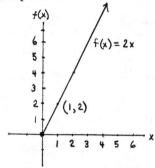

Domain of f
= set of positive real numbers

Function g: g(x) = 2x

g(x) is the number of children in x sets of twins.

Graph of g:

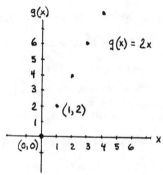

Domain of g
= set of nonnegative integers

If you let y = 2x in the examples, you would not be able to distinguish the functions. In this way, function language is sometimes clearer.

Here is another way of thinking of domains. The domain of a function is the replacement set for the first variable. When no domain is given, the domain is assumed to be the largest set possible.

Examples: Domains of Functions

1. The function $\{(2, 20), (4, 45), (7, -3)\}$ has domain $\{2, 4, 7\}$.

2. If $A(t)$ is the area of a triangle t, then the domain of A is the set of all triangles.

3. A function is given with equation $f(x) = 3x - 2$. No domain is given. Then the domain is assumed to be the set of real numbers. That is, x can be any real number.

Questions covering the reading

1. Define: linear function.

2. Every non-vertical line has an equation of the form _____.

3. Every linear function has an equation of the type _____.

4. The graph of $f(x) = \frac{1}{2}x - 8$ is a _____ with slope _____ and y-intercept _____.

5. Give the slope of the line with equation $g(x) = 10x + 3$.

6. The slope determined by (x_1, y_1) and (x_2, y_2) is _____.

7. Using function notation, what is the formula for slope?

8. Define: domain of a function.

9. Draw graphs of two functions with different domains which have the equation $y = 3x$.

10. Why is function language sometimes clearer?

11-16. What is the domain of each function?

11. $\{(0, -6), (2, 7), (11, 6)\}$ 12. $\{(Abe, 206), (Grace, 123)\}$

13. f where $f(x) = 2x - 6$ and x can be any integer.

14. g where $g(x) = x^2$ and $-3 < x < 3$.

15. A where $A(t)$ is the area of a triangle t.

16. h where $h(x) = 3x$.

17-20. (a) Graph the function if the domain is $\{0, 1, 2, 3, 4\}$.
(b) Graph the function if the domain is the set of real numbers.

17. $f(x) = 3x - 5$ 18. $g(x) = \frac{2}{3}x + 1$

19. linear function with slope 1 and y-intercept 2.

20. linear function with slope -1 and y-intercept 3.

Questions extending the reading

1-8. The <u>range</u> of a function is the set of all <u>second</u> coordinates of a function. Give the range for each function.

1. $\{(1, 2),\ (3, 5),\ (8, 13)\}$ 2. $\{(\text{Carol},\ 15),\ (\text{Mike},\ 17)\}$

3. function f graphed on page 701

4. function g graphed on page 701

5. the function f, equation $f(x) = 3x + 6$

6. the function g, equation $g(x) = x^2$

7. the function h, equation $h(n) = 3n$, where the domain of h is $\{2, 4, 6\}$.

8. the function s, equation $s(t) = 3t$, where the domain of s is the set of real numbers.

9-11. Refer to Questions 1-8. A <u>constant function</u> is a function with only one element in its range.

9. Graph $f(x) = 3$, one example of a constant function.

10. The graph of a constant function is all or part of a _____ line.

11. The slope of the graph of a constant function is _____.

703

12-15. Find an equation for each linear function.

12. C, where C(n) is the cost of n 13¢ stamps.

13. P, where P(s) is the perimeter of a square with side s.

14. L, where L(d) is what is left from $50 after purchasing d doorknobs at $6.95 each.

15. C, where C(m) is the cost of renting a car for a day at $18 per day plus 13¢ a mile.

Lesson 4

Absolute Value Functions

The function with equation

$$f(x) = |x|$$

is the simplest example of an absolute value function. By substitution, you can find ordered pairs for this function.

If x = 2, f(x) = |x| = |2| = 2.   Pair (2, 2)

If x = -8, f(x) = |x| = |-8| = 8.   Pair (-8, 8)

If x = 0, f(x) = |x| = |0| = 0.   Pair (0, 0)

All ordered pairs of this function are graphed below.

f(x) = |x| The graph of this
                             function is an angle.

You already know many uses for lines; these are all uses of

linear functions.   But you may wonder why you would want to study

such "weird" functions as absolute value functions.   One reason is

705

that these functions <u>are</u> found in many applications. A second reason
is that complicated situations can sometimes be better understood if
you know some simple functions.

Example: A plane crosses a check point going due east. It flies

       at 600 km/hr for 2 hours, then returns flying due west.
       Is there any formula describing its distance from the check
       point?

Solution: Let d(t) = its distance from the check point t hours after
       crossing. The given information tells you some values of
       the function.

       When t = 0, it is at the check point.   So d(0) = 0.

       When t = 1, it is 600 miles east.   So d(1) = 600.

       When t = 2, it is 1200 miles east.   So d(2) = 1200.

       When t = 3, it is 600 miles east.   So d(3) = 600.

       When t = 4, it is again at the check point. So d(4) = 0.

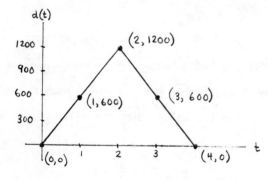

When all pairs are
graphed, the graph
at left results.

706

The graph is part of an angle. So you should expect a formula

for d(t) to involve the absolute value symbol. This is exactly what

happens.

$$d(t) = -600|t - 2| + 1200$$

In your next algebra course, you may learn how you could find this

formula. Without knowing the graph of f(x) = |x| you would never

think of using absolute value.

In earlier chapters, you would have written $d = -600|t - 2| + 1200$.

Using the d(t) function notation shows that the value of d depends on

the time. There may be other things which depend on the time. It

is common to see things like:

f(t) = amount of fuel used t hours after crossing

a(t) = altitude t hours after crossing

etc.

In this way, it is clear that the fuel, altitude, etc., all depend on the

time.

Questions covering the reading

1. Give an example of an absolute value function.

2. The graph of an absolute value function is _____.

3-6. Let f(x) = |x|. Calculate

3. f(3)   4. f(-2)   5. f(0)   6. $f(-\frac{1}{2})$

7. Graph the function f in Questions 3-6.

8. Name two reasons for studying absolute value functions.

9. Name two reasons why the f(x) function notation is helpful. (One reason is given in this lesson; one is in the previous lesson.)

10-15. Suppose $d(t) = -600 |t - 2| + 1200$.

10. Calculate d(0).   11. Calculate d(3).

12. Calculate d(2.5).   13. Calculate d(1.2).

14. Describe a situation which can lead to this function.

15. In the example of this lesson, what is the domain of d? How was this domain determined?

Questions testing understanding of the reading

1-2. Refer to the example of this lesson, where
$$d(t) = -600 |t - 2| + 1200.$$

1. Although -1 is not in the domain of d, it is possible to calculate d(-1). Do this. What does your answer mean?

2. Repeat Question 1 for d(6).

3-4. Graph the function with the given equation.

3. $f(x) = |2x|$ 4. $y = |x - 3| + 4$

5-7. Start at the goal line of a football field. Walk to the other goal line. After you have walked w yards, you will be on the y yard line.

5. Multiple choice. Which equation relates w and y?
   (a) $y = w$ (b) $y = |w|$
   (c) $y = |50 - w| + w$ (d) $y = -|w - 50| + 50$

6. Let $y = f(w)$. What is the domain of f?

7. Let $y = f(w)$. Graph f.

Lesson 5

Quadratic and Other Functions

Let $y = f(x)$. A <u>quadratic function</u> f is when x and y are related so that

$$dxy + ey = ax^2 + bx + c, \quad \text{and} \quad de \neq 0.$$

This is quite complicated in general. But you have seen two simple examples of quadratic functions.

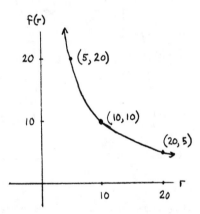

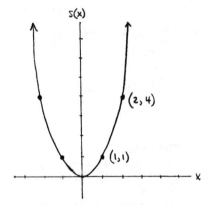

rate and time to travel 100 miles (page 395)

$$f(r) = \frac{100}{r}$$

Domain = set of positive reals

The curve is one branch of a <u>hyperbola</u>.

squaring function (page 695)

$$s(x) = x^2$$

Domain = set of all reals

The curve is a <u>parabola</u>.

709

All quadratic functions have graphs which are either hyperbolas or parabolas. Here is a function in which the equation makes it obvious that it should be a quadratic function.

$$f(x) = -x^2 + 4x + 5$$

Some ordered pairs:

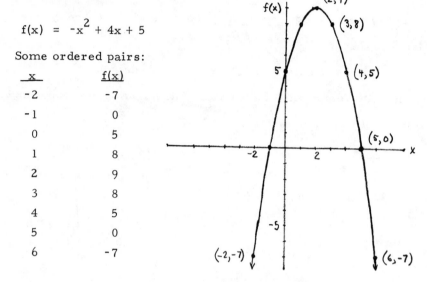

x	f(x)
-2	-7
-1	0
0	5
1	8
2	9
3	8
4	5
5	0
6	-7

The paths of baseballs or other objects thrown into the air are in the shape of parabolas like this one.

Many other curves can be the graphs of functions. The equation $y = 2^x$ describes an exponential function you have seen before. An exponential function is one where the domain variable is found only in the exponent.

710

$$f(x) = 2^x$$

Some points of f are:

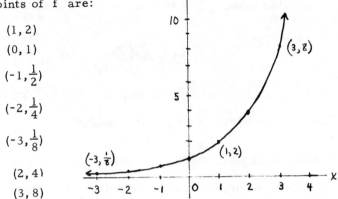

(1, 2)

(0, 1)

$(-1, \frac{1}{2})$

$(-2, \frac{1}{4})$

$(-3, \frac{1}{8})$

(2, 4)

(3, 8)

Exponential functions describe many types of growth patterns.

Questions covering the reading

1. Give an example of an equation for a quadratic function.

2. The graph of a quadratic function may either be a _____ or a _____.

3-8. Graph the function with the given equation.

3. $f(x) = \frac{1}{x}$

4. $y = x^2$

5. $y = -x^2$

6. $g(x) = -x^2 + 6x + 2$

7. $h(x) = 2^x$

8. $y = 3^x$

9. The path of an object thrown into the air is in the shape of a
_____.

10. Let r be your average rate in km per hour, $r \leqslant 200$. Let $f(r)$ be the time it takes to go 1000 km. Graph f.

1-4. <u>Multiple choice</u>. Pick the type of function corresponding to each situation.

                    (a) linear
                    (b) quadratic
                    (c) exponential

1. The cost of potatoes depends upon how much you buy.

2. The average speed of a car determines how far the car will go in an hour.

3. Thickness of glass determines how much light gets through.

4. The average speed of a car determines how long it will take to get home.

5-8. Graph the function with the given equation.

5. $y = -2x^2 + 3x$ 6. $f(x) = \dfrac{1}{x + 2}$

7. $g(x) = (\frac{1}{2})^x$ 8. $y = x^2 - 10$

9-10. Linear and some quadratic functions are special types of <u>polynomial</u> functions. Graph these functions.

9. f, where $f(x) = x^3 - 3x^2$

10. g, where $g(n) = 2n^3 - 6n^2 - 10$

11-12. Graph these <u>power</u> functions.

11. $y = x^3$ 12. $y = x^4$

<u>Question for exploration</u>

1. In a dark room, hold a flashlight parallel to a wall. Turn on the flashlight and look at the outline of the light on <u>that</u> wall. What curve is outlined?

Lesson 6

Probability Functions

Throw two fair dice. Then add the numbers which appear on the dice. Let P(n) = the probability that the sum is n. P is one kind of probability function.

The domain of a probability function is the set of events in an experiment. The values of the function are their probabilities. The only possible values of n are $2, 3, 4, \ldots, 12$. So the domain is $\{2, 3, 4, \ldots, 12\}$. You calculated the values of this function in Chapter 13. Here they are again.

n	P(n)
2	1/36
3	2/36
4	3/36
5	4/36
6	5/36
7	6/36
8	5/36
9	4/36
10	3/36
11	2/36
12	1/36

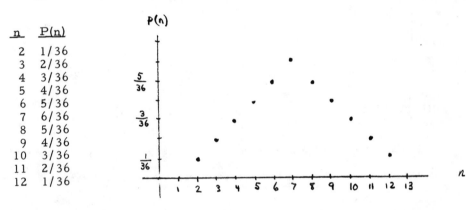

Since the values of P(n) add to 1, there are no more possibilities.

The graph is part of an angle. This suggests that there is a formula for P(n) which uses the $|\quad|$ sign. Here it is.

$$P(n) = -\frac{1}{36}\left|n - 7\right| + \frac{1}{6}$$

713

That formula has no meaning except when n is in the domain of the function.

Here are two simpler examples of probability functions. Suppose a coin is tossed. At left is the function if P(heads) = $\frac{1}{2}$. At right is the function if P(heads) = .6.

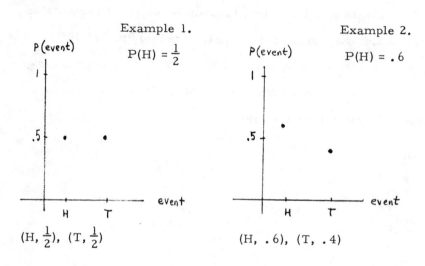

Example 1.

P(H) = $\frac{1}{2}$

(H, $\frac{1}{2}$), (T, $\frac{1}{2}$)

Example 2.

P(H) = .6

(H, .6), (T, .4)

In Example 1, both events have the same probability. The events occur randomly. Here is another example with random events.

The circle is divided into 5 con-gruent regions. Assume all regions have the same probability of the spinner landing in them.

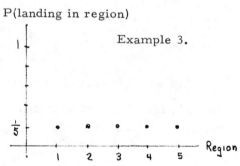

P(landing in region)

Example 3.

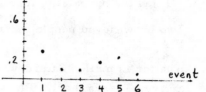

1-3. At right is graphed a probability function for a weighted 6-sided die.

1. $P(1) \approx$ _____.

2. $P(6) \approx$ _____.

3. How would the graph look if the die were fair?

4. What is the domain of a probability function?

5. What are the values of a probability function?

6. Why can't 2 be a value of a probability function?

7-12. Refer to the dice experiment, page 713.

7. A sum of _____ is as likely as a sum of 9.

8. A sum of _____ is most likely.

9. A sum of _____ is least likely.

10. If $P(t) = \frac{4}{36}$, then t = _____ or _____.

11. If $P(x) = \frac{1}{36}$, then x = _____ or _____.

12. If $P(n) = -\frac{1}{36} \left| n - 7 \right| + \frac{1}{6}$, calculate $P(8)$. What have you calculated?

13-15. Graph the probability function suggested by each situation.

13. A fair coin is tossed.

14. A coin is tossed and $P(\text{heads}) = \frac{3}{8}$.

15. The spinner pictured at right is spun.

1-2. Graph the probability function suggested by each situation.

1. Two brown-eyed people marry and P(child will be brown-eyed) = $\frac{3}{4}$.

2. A letter is mailed Saturday. P(it arrives Monday) = $\frac{1}{2}$. P(it arrives Tuesday) = $\frac{1}{4}$. P(it arrives Wednesday) = $\frac{1}{8}$. P(it arrives Thursday) = $\frac{1}{16}$. P(it arrives Friday) = $\frac{1}{32}$. P(it arrives Saturday) = $\frac{1}{64}$. Calculate P(it never arrives) and graph the appropriate probability function.

3. For example 3, page 714, what is a formula for P(n)?

4. Let H = 1, T = 2. Then for Example 1, page 714, which is a formula for P(n)?

   (a) P(n) = 1   (b) P(n) = 2

   (c) P(n) = $\frac{1}{2}$ (d) none of these

5. Let H = 1, T = 2. Then for Example 2, page 714, which is a formula for P(n)?

   (a) P(n) = .8 - .2n   (b) P(n) = .6n

   (c) P(n) = .5n + 1   (d) none of these

6. If a fair 6-sided die is tossed, then what is a formula for P(n)?

7-12. Suppose three fair coins are tossed at the same time. There are then 8 possible outcomes: HHH, HHT, HTH, HTT, THH, THT, TTH, TTT. Let P(n) = the probability that n heads occur. Calculate:

7. P(0)   8. P(1)   9. P(2)   10. P(3)   11. P(4)

12. Graph the function P from Questions 7-11.

13-18. Repeat Questions 7-12 if four fair coins are tossed at the same time.

Frequency Distributions

To find out the ages of algebra students, you can sample. In a sample done by the author in one class of 25 students, ages ranged from 13 to 17. The frequencies are summarized and graphed here.

Age	Frequency
13	2
14	8
15	9
16	3
17	3

Frequency distribution (1)

The set of ordered pairs (event, frequency of event) is called a frequency distribution. A frequency distribution is a function which relates events to their frequencies.

Now suppose you want to know how much these students spend for lunch. Here there are lots of outcomes--one for each price. You would expect each frequency to be small.

In this kind of situation, one way to summarize the data is to group outcomes of the same type. When the outcomes are numerical, intervals are picked. (An interval is the set of numbers between two numbers, possibly including the numbers.) Normally the intervals have the same length and each event studied corresponds to an interval.

Here is a summary of what 25 students spent for lunch.

Interval	0-39¢	40-79¢	$.80-1.19	$1.20-1.59	$1.60-1.99
Frequency	8	9	4	3	1

When intervals are used for events, the graph of the frequency distribution often uses rectangles instead of points. This kind of graph is called a <u>histogram</u>.

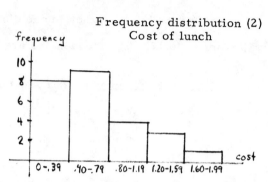

Frequency distribution (2)
Cost of lunch

Frequency distribution (2) shows that more students spent between 40¢ and 79¢ for lunch than in any other interval. This interval is called the <u>mode</u> of the distribution. In the age distribution of p. 717, the mode is 15.

Definition:

> The <u>mode</u> of a distribution is the event which has the highest frequency. It is the event which occurs most often.

<u>Questions covering the reading</u>

1-5. What does each term or phrase mean?

1. frequency of an event   2. frequency distribution

3. interval   4. histogram

5. mode of a frequency distribution

718

6. A frequency distribution is a function which relates _____ to _____.

7-10. Graphed at right is a frequency distribution of 100 numbers. It possibly could represent a distribution of IQ's.

7. What is the frequency of the event 55-69?

8. What is the frequency of the event 115-129?

9. What is the mode?

10. Name the event with the least frequency.

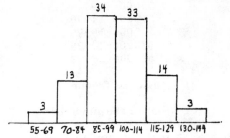

11. Answer Questions 9 and 10 for frequency distribution (1) of this lesson.

12. Answer Questions 9 and 10 for frequency distribution (2) of this lesson.

Questions testing understanding of the reading

1-4. Multiple choice. Pick the best answer from these choices.

(a) set of real numbers

(b) set of real numbers between 0 and 1

(c) set of events

(d) set of natural numbers

1. The domain of a frequency distribution is a _____.

2. The values of a frequency distribution are from the _____.

3. The domain of the probability function is a _____.

4. The values of a probability function are from the _____.

5. In a certain community, a count was made of the incomes of 6060 families. Results were as follows:

Income	Frequency
under $5000	297
$5000-9999	1060
$10000-14999	2606
$15000-19999	1100
$20000-24999	759
$25000-29999	238

(a) Graph the frequency distribution.

(b) What is the mode of this distribution?

6. On a test, 20 students scored as follows:

Interval	50-59	60-69	70-79	80-89	90-100
Frequency	1	0	5	7	7

(a) Graph the frequency distribution.

(b) What is the mode of this distribution?

7. Toss a 6-sided die 50 times.

(a) Graph an appropriate frequency distribution.

(b) What is the mode of your distribution?

(c) Do you believe that your die is fair?

8. Repeat Question 7 but for a coin which is tossed 50 times.

9-13. Suppose you had to graph a frequency distribution using the given type of data. Would you want to use a histogram?

9. for 30 people, the number of days to build an amplifier from a kit

10. 100 estimates of the height of a building

11. the family size in each of 1000 families

12. speeds of 50 typists

13. results of 75 spins of the spinner at right

14. Why should the intervals in a histogram be nearly the same length?

15. Draw a histogram in which the intervals are not the same length. Give an example to show that this kind of histogram can be misleading.

16. A histogram is a special type of

    (a) circle graph   (b) line graph   (c) bar graph

17-18. The mode of a sequence of scores is the most common score. (That is, again it is the score which occurs with the most frequency.) Give the mean and mode for each sequence.

17. 80, 70, 70, 70, 80, 90   18. 1, 2, 1, 3, 1, 4, 1, 5

19. Let $f(n)$ = "frequency of students with age n." Referring to frequency distribution (1) in this lesson, calculate:

    (a) $f(12)$ (b) $f(13)$ (c) $f(16)$ (d) $f(14)$

20. Refer to Question 18 and let $s(n)$ = frequency of score n. Calculate:

    (a) $s(1)$ (b) $s(2)$ (c) $s(3)$ (d) $s(5)$

Questions for exploration

1. Look in a newspaper or a magazine for a histogram. (a) Do you think the histogram accurately pictures the data? (b) Could the data have been presented without using a histogram?

2. Some distributions are called bimodal. What do you think this word means? Look in a dictionary to check your thinking. What kinds of situations might lead to bimodal distributions?

3. Make up your own situation from which you could get a frequency distribution. If you have time, collect the data and graph the distribution.

Lesson 8

The Chi-Square Statistic

The first ideas from statistics in this course were frequencies and sampling. You studied these ideas early in Chapter 1. In this, the last lesson of this course, we return to those ideas by considering an amazing statistic. This statistic compares sample frequencies with expected frequencies. It is one of the most often-used of all statistics. Here is a typical use.

In a large factory, over a two-year interval, there were 180 accidents. Someone noticed that more accidents seemed to occur on Mondays. So the data was split up by day of the week. Does Monday seem to be special?

Observed (Sample) Frequencies: Frequency distribution o					
Day of the week	M	T	W	T	F
Number of accidents	55	38	23	22	42

If accidents occurred randomly, the expected frequency of accidents for each day would be $\frac{180}{5}$ or 36.

Expected Frequencies: Frequency distribution e					
Day of the week	M	T	W	T	F
Number of accidents	36	36	36	36	36

Are the accidents occurring randomly?

To answer this question, we want to know: How often would random events deviate from e as much as o is deviated from e? If this would happen less than .05 of the time, the event is unusual. If this would happen less than .01 of the time, the event is highly unusual. Then randomness would be strongly questioned. People in the factory would then look for reasons why there are so many accidents on Monday and so few on Wednesday or Thursday, so that the situation could be corrected.

In 1900, the English statistician Karl Pearson developed a way of measuring the deviation of frequencies in o from the frequencies in e. It uses a number called the <u>chi-square</u> statistic. ("Chi" is pronounced "ky" as in "sky.")

<u>Algorithm for calculating the Chi-square statistic</u>:

<u>Step 1.</u> Count the number of events. Call this number n.
(Example 1: In our case, there are 5 events, one each for M, T, W, T, F. So n = 5.)

<u>Step 2.</u> Let o_1, o_2, ..., o_n be the observed frequencies.
($o_1 = 55$ $o_2 = 38$ $o_3 = 23$ $o_4 = 22$ $o_5 = 42$)

<u>Step 3.</u> Let e_1, e_2, ..., e_n be the expected frequencies.
($e_1 = 36$ $e_2 = 36$ $e_3 = 36$ $e_4 = 36$ $e_5 = 36$)

Step 4. Calculate $\dfrac{(o_1 - e_1)^2}{e_1}$, $\dfrac{(o_2 - e_2)^2}{e_2}$, ..., $\dfrac{(o_n - e_n)^2}{e_n}$.

$$\left(\begin{array}{ll} \dfrac{(o_1 - e_1)^2}{e_1} = \dfrac{(55-36)^2}{36} = \dfrac{361}{36} & \dfrac{(o_2 - e_2)^2}{e_2} = \dfrac{(38-36)^2}{36} = \dfrac{4}{36} \\[3ex] \dfrac{(o_3 - e_3)^2}{e_3} = \dfrac{(23-36)^2}{36} = \dfrac{169}{36} & \dfrac{(o_4 - e_4)^2}{e_4} = \dfrac{(22-36)^2}{36} = \dfrac{196}{36} \\[3ex] \dfrac{(o_5 - e_5)^2}{e_5} = \dfrac{(42-36)^2}{36} = \dfrac{36}{36} & \end{array} \right)$$

Step 5. Sum the n numbers found in Step 4. This sum is the chi-square statistic.

$$\left(\dfrac{361}{36} + \dfrac{4}{36} + \dfrac{169}{36} + \dfrac{196}{36} + \dfrac{36}{36} = \dfrac{766}{36} \approx 21.2 \right)$$

The chi-square statistic measures how far observed scores are deviated from the expected scores. The larger it is, the greater the deviation. But is 21.2 unusually large? You can find out by looking at the table on page 725. $n - 1 = 4$. That row shows that a value $\geqslant$ 18.5 would occur only with probability .001. So we have evidence that the accidents are _not_ evenly distributed among days. The people in the factory should try to figure out why Monday has more accidents.

TABLE: Critical Chi-square Values

	Probability			
n-1*	.10	.05	.01	.001
1	2.71	3.84	6.63	10.8
2	4.61	5.99	9.21	13.8
3	6.25	7.81	11.34	16.3
4	7.78	9.49	13.28	18.5
5	9.24	11.07	15.09	20.5
6	10.6	12.6	16.8	22.5
7	12.0	14.1	18.5	24.3
8	13.4	15.5	20.1	26.1
9	14.7	16.9	21.7	27.9
10	16.0	18.3	23.2	29.6
15	22.3	25.0	30.6	37.7
20	28.4	31.4	37.6	45.3
25	34.4	37.7	44.3	52.6
30	40.3	43.8	50.9	59.7
50	63.2	67.5	76.2	86.7

* n is the number of events.

How to read the table: The number 14.1 appears in column .05, row 7. This means that, with 8 events, a chi-square value greater than 14.1 occurs with probability .05 or less.

How to use the table: Suppose you have 10 events and calculate a chi-square value of 20.38. Since n = 10, n-1 = 9. So you look at row 9. A chi-square value as large as 16.9 only occurs with probability of .05. Since 20.38 > 16.9, you should question the way you calculated the expected frequencies. (It would take a chi-square value of 21.7 to strongly question.)

Example 2: Two coins are tossed 100 times. 0 heads occurred 20 times; 1 head occurred 54 times, 2 heads occurred 26 times. If the coins are fair, $P(0 \text{ heads}) = \frac{1}{4}$, $P(1 \text{ head}) = \frac{1}{2}$, and $P(2 \text{ heads}) = \frac{1}{4}$. So we expect 25, 50, and 25 as frequencies. Does the tossing indicate that the coins are not fair?

Solution: You might organize your work as follows:

Event	observed	expected	o-e	$(o-e)^2$	$\frac{(o-e)^2}{e}$
0 heads	20	25	-5	25	$\frac{25}{25} = 1$
1 head	54	50	4	16	$\frac{16}{50} = .32$
2 heads	26	25	1	1	$\frac{1}{25} = .04$

chi-square value = Sum = 1.36

Since n = 3, n - 1 = 2. Look at row 2 of the table. A sum as large as 4.61 would occur 10% of the time. So 1.36 is very common and the distribution is not at all unusual.

Notes about the chi-square statistic: (1) The chi-square statistic can be used whenever there are observed frequencies and you have some way of calculating expected frequencies. (2) The chi-square value is not as good a measure of the deviation when there is an expected frequency which is less than 5. (3) The mathematics needed to calculate the values in the table is beyond that normally studied in high school. If you study a full course devoted to statistics, you may use further applications of the chi-square statistic.

Questions covering the reading

1. What does the chi-square statistic measure?

2-6. With the given chi-square value and number of events, should you (a) not question, (b) question, or (c) strongly question the assumed probabilities which have led to the expected frequencies? (Of course, you need to use the table on p. 725.)

2. 6 events, chi-square value = 11.58

3. 16 events, chi-square value = $40\frac{1}{2}$

4. 2 events, chi-square value = 5

5. 7 events, chi-square value = 14.02

6. 5 events, chi-square value = 13.5

7. Suppose you toss a coin 100 times and heads occurs 63 times. (Tails occurs 37 times.) You wish to test whether the coin is fair.

    (a) What are the expected frequencies?

    (b) What are the observed frequencies?

C   (c) Calculate the chi-square statistic for this experiment.

    (d) What value of n should be used for the table?

    (e) What row of the table do you look in?

    (f) Is there evidence to believe the coin is not fair?

8. Repeat Question 7 for the following situation: Two coins are thrown 100 times with the following results: HH occurred 20 times, HT or TH occurred 60 times, TT occurred 20 times.

9. The chi-square statistic was developed in the year _____ by _____.

10-13. (a) Find the value of the chi-square statistic for the following observed frequencies of accidents in a 5-day week. (b) Is there evidence to believe that the accidents are not occurring randomly?

	M	T	W	T	F
10.	20	20	20	20	20
11.	22	18	20	17	23
12.	23	22	20	18	17
13.	25	15	15	30	15

14-19. Two coins are tossed 100 times. Results are given. Is there reason to believe the coins are not fair? (Use the chi-square statistic to answer.)

	2 heads	1 head, 1 tail	2 tails
14.	30	48	22
15.	23	55	22
16.	25	40	35
17.	29	52	19
18.	20	70	10
19.	25	75	0

20. When can the chi-square statistic be used?

Questions applying the reading

1-4. Answer these questions by using the chi-square statistic. (You must begin by calculating expected frequencies.)

1. Testing a die. In 60 tosses of a die, the following frequencies of outcomes are found.

Outcome	1	2	3	4	5	6
Frequency	13	9	9	15	6	8

Is there evidence that the die is not fair?

2. <u>Testing a spinner.</u> You build a spinner as at right and spin it 40 times.

Outcome	1	2	3	4	5
Frequency	10	6	4	6	14

Do you think the spinner is fair?

C 3. <u>Do earthquakes occur more in some seasons?</u> <u>The Official Associated Press Almanac</u> for 1975 lists 39 earthquake disasters since 1925. Here are their frequencies by month:

Jan	Feb	Mar	Apr	May	Jun	Jul	Aug	Sep	Oct	Nov	Dec
2	3	6	2	5	2	5	5	2	1	0	6

Since the expected frequency 39/12 is less than 5, a chi-square procedure is not appropriate. So we group the data two months at a time. Now the expected frequency is 39/6 or 6.5.

J-F	M-A	M-J	J-A	S-O	N-D
5	8	7	10	3	6

Does the data support a view that more earthquakes occur at certain times of the year?

C 4. <u>Do baseball teams score more in the beginning, middle, or end of a game?</u> Here are the total runs scored in each inning from the 14 Major League baseball games of April 14, 1975.

Innings	1	2	3	4	5	6	7	8	9
Runs	10	7	16	19	23	5	9	7	9

Group innings 1-3, 4-6, and 7-9 together. Does the information lead you to believe that teams score more runs in the middle?

5. <u>Assume a coin is fair.</u> Calculate the chi-square values for (a) 6 heads in 10 tosses, (b) 60 heads in 100 tosses, and (c) 600 heads in 1000 tosses. Which event is most likely?

6. <u>Were the people guessing?</u> Sixty people were asked to name the U.S. President in 1850 from the names below.

14 picked Millard Fillmore
25 picked Abraham Lincoln
21 picked Martin Van Buren

(a) Is there evidence to believe the people were just guessing? (Use the chi-square statistic to answer this.)

(b) Who was President in 1850?

C 7. <u>School spirit</u>. A high school has 260 freshmen, 255 sophomores, 245 juniors, and 240 seniors. Signed up for a bus are 12 freshmen, 6 sophomores, 8 juniors, and 14 seniors. Is there evidence to believe that the classes do not have equal school spirit?

8. <u>Testing aspirin</u>. 100 people were given two different types of aspirin and asked to record which acted faster. Type A (plain aspirin) acted faster than type B (buffered aspirin) in 41 of 100 cases. Is the evidence strong enough to think that type A is slower-acting?

9-10. You ask 40 people whether they prefer candidate A or candidate B. You wish to test whether the election is even. Suppose x people prefer A. Then the chi-square value is

$$\frac{(20 - x)^2}{10}.$$

C 9. (a) What values of x give a chi-square value of 3.84?

(b) What values of x give a chi-square value of greater than 3.84? (These values indicate an unusual event.)

C 10. (a) What values of x give a chi-square value of 6.63?

(b) What values of x give a chi-square value of greater than 6.63? (These values indicate a highly unusual event.)

C 11-12. For a fair coin, the chi-square value for h heads in 100 tosses is $\frac{(50 - h)^2}{25}$.

11. What values of h give a chi-square value which is greater than 3.84? (These values indicate an unusual event.)

12. In this situation, there is a .01 probability of a chi-square value greater than 6.63. What values of h give such a highly unusual chi-square value?

13. Testing astrology. From Who's Who in American Politics, 5th Edition, 1975-76 (R.P. Bowker, publishers), 96 women were picked by opening the book every few pages and choosing the first woman listed on the page. (This is a scatter process and is not random but is close to random.) Birthdates were recorded. The births fell under the usual birth signs with the following frequencies:

Aries	6	Leo	12	Sagittarius	11
Taurus	8	Virgo	5	Capricorn	7
Gemini	7	Libra	12	Aquarius	8
Cancer	8	Scorpio	4	Pisces	8

Does this data lead you to believe that certain birth signs are more likely to produce women politicians? (Hint: Calculate the chi-square statistic assuming randomness is expected; then use row 10 from the chi-square table.)

Questions for discussion and exploration

1. Repeat Question 4, p. 729, with more recent data.

2. Do a second test of birth signs as in Question 13 (above).

3. Make up your own situation to test using the chi-square statistic.

Chapter Summary

A _function_ is a set of ordered pairs in which each first coordinate appears with exactly one second coordinate. A function may be described by a graph, by a written rule, by listing the pairs, or by an equation. If you know the description for a function, then knowing the first coordinate of a pair is enough knowledge to determine the second coordinate of that pair. For this reason, functions exist whenever one thing determines another.

Sets of ordered pairs are common in mathematics and can often be graphed. If a function f contains the ordered pair (a, b), then you can write f(a) = b. The set of possible values of a is the _domain_ of the function. b is a _value_ of the function.

Many of the ideas in earlier chapters are examples of functions. Most lines are graphs of _linear functions_. Some angles are graphs of _absolute value functions_. Quadratic expressions lead to _quadratic functions_; their graphs are hyperbolas or parabolas. The equation $f(x) = 2^x$ describes an _exponential function_. _Probability functions_ are functions which relate events to their probabilities. _Frequency_

732

<u>distributions</u> are functions which relate events to their frequencies. The chi-square statistic is a statistic which measures how close one function is to another.

Look back at the table of contents of this book. You will see that almost all of the types of expressions and major concepts are found in this chapter. Functions are everywhere, both in the mathematics and in its applications. If you study more algebra, you are certain to study functions in more detail. In this way, the end of this book provides a beginning for later work.

REFERENCE INFORMATION

MODELS FOR OPERATIONS

A model for an operation is a general pattern which includes many of the uses of the operation. Each fundamental operation has a model corresponding to each of the most important uses of numbers. Repetition models relate the operations to each other.

Operation	Counting	Measurement	Comparison or Change	Repetition
Addition	union	joining	slide	
Subtraction	take-away	cutting-off	directed distance	
Multiplication	ordered pair	area	size change	repeated addition
Division	splitting-up	rate	scale comparison	repeated subtraction
Powering	permutation		growth	repeated multiplication

POSTULATES FOR REAL NUMBER OPERATIONS

A <u>postulate</u> is a statement which is assumed to be true. The postulates for real number operations are suggested by the models for the operations. They also can be verified by the models. Let a, b, and c be any real numbers. Then:

Type of Postulate:	Name of Property	Addition	Multiplication
Existence:	Closure	$\underline{a + b}$ is a real number.	$\underline{ab}$ is a real number
	Identity	There is a unique real number 0 with $a + 0 = a$	There is a unique real number 1 with $a \cdot 1 = a$
	Inverse	There is a unique real number $-a$ with $a + -a = 0$	When $a \neq 0$, there is a unique real number $\frac{1}{a}$ with $a \cdot \frac{1}{a} = 1$
Assemblage:	Commutative	$a + b = b + a$	$ab = ba$
	Associative	$a + (b + c) = (a + b) + c$	$a(bc) = (ab)c$
Sentence-Solving:	Equations	If $a = b$, then $a + c = b + c$.	If $a = b$, then $ac = bc$.
	Inequality	If $a < b$, then $a + c < b + c$.	If $a < b$ and $c > 0$, then $ac < bc$. If $a < b$ and $c < 0$, then $ac > bc$.
	Distributive	$ac + bc = (a + b)c$	
	Power	$c^a \cdot c^b = c^{a+b}$	

ALGORITHMS FOR SENTENCE-SOLVING

An _algorithm_ is a procedure for solving a particular kind of problem. All algorithms for solving sentences follow from the postulates for real number operations.

Sentence (solving for x)	Algorithm
Type 1: $ax = b$ $ax < b$	Multiply both sides by $\frac{1}{a}$.
Type 2: $a + x = b$ $a + x < b$	Add $-a$ to each side.
Type 3: Proportion	Apply the Means-Extremes Property. The resulting sentence is of Type 1.
Type 4: $ax + b = c$ $ax + b < c$	Add $-b$ to each side. The resulting sentence is of Type 1.
Type 5: $ax + b = cx + d$ $ax + b < cx + d$	Add $-cx$ to each side. The resulting sentence is of Type 4.
Type 6: $ax^2 + bx + c = 0$	Use the Quadratic Formula.

There are four algorithms used in solving linear systems. They are (1) substitution, (2) graphing, (3) addition or subtraction, and (4) multiplication followed by addition or subtraction.

In all sentence-solving, trial and error is a possible algorithm. But trial and error is risky.

THEOREMS ABOUT REAL NUMBER OPERATIONS

A <u>theorem</u> is a statement which follows from one or more postulates. There are infinitely many theorems about operations on real numbers. Here are the most important. They are grouped by type. For all real numbers a, b, B, c, d, k, m, n, x, and y:

<u>Theorems about Zero</u>:

Zero Multiple:	$0 \cdot x = 0$
Zero Power:	$x^0 = 1$
Zero Product:	If $ab = 0$, then $a = 0$ or $b = 0$.

<u>Theorems about Opposites</u>:

Opposite of a Sum:	$-(a + b) = {}^-a + {}^-b$
Oppositing Property of $^-1$:	$^-1 \cdot x = {}^-x$
Multiplication:	$^-x \cdot {}^-y = xy \qquad x \cdot {}^-y = {}^-xy$
Division:	$\dfrac{^-x}{^-y} = \dfrac{x}{y} \qquad \dfrac{x}{^-y} = \dfrac{^-x}{y} = \dfrac{x}{y}$

<u>Theorems from Distributivity</u>:

with Subtraction:	$(a - b)x = ax - bx$
Commutativity variant:	$c(a + b) = ca + cb$

<u>Theorems about Fractions</u>:

Simplification:	$\dfrac{ka}{kb} = \dfrac{a}{b}$
Addition:	$\dfrac{a}{c} + \dfrac{b}{c} = \dfrac{a + b}{c}$
Multiplication:	$\dfrac{a}{b} \cdot \dfrac{c}{d} = \dfrac{ac}{bd}$

739

<u>Theorems for Sentence-Solving</u>:

Means-Extremes Product: If $\frac{a}{b} = \frac{c}{d}$, then $ad = bc$.

Quadratic Formula: If $ax^2 + bx + c = 0$, then

$$x = \frac{-b \pm \sqrt{b^2 - 4ac}}{2a}$$

<u>Theorems about Powers</u>:

Negative Exponent: B^{-n} is the reciprocal of B^n.

Division: $\dfrac{B^m}{B^n} = B^{m-n}$

Power of a Power: $(B^m)^n = B^{mn}$

Power of a Product: $(xy)^n = x^n y^n$

Power of a Quotient: $(\frac{x}{y})^n = \dfrac{x^n}{y^n}$

<u>Theorems about Square Roots</u>:

Power: $B^{\frac{1}{2}} = \sqrt{B}$

Product: $\sqrt{xy} = \sqrt{x} \cdot \sqrt{y}$

Quotient: $\sqrt{\dfrac{x}{y}} = \dfrac{\sqrt{x}}{\sqrt{y}}$

<u>Theorems about Polynomials</u>:

FOIL: $(a + b)(c + d) = ac + ad + bc + bd$

Polynomial Pattern: Given a sequence, take differences of consecutive terms. If the differences are equal the nth time this is done, then s_n is a polynomial of degree n.

<u>Theorems about Applications of Square Roots</u>:

Growth: If a quantity is multiplied by B in a given period, then in $1/2$ the time (at the same rate) the quantity is multiplied by $\sqrt{B}$.

Middle of Growth: If a quantity grows exponentially with $\underline{a}$ in the beginning and $\underline{b}$ at the end, then in the middle of this period there is $\sqrt{ab}$.

Pythagorean: If a, b, and c are lengths of sides of a right triangle and c is the length of the longest side, then

$$c = \sqrt{a^2 + b^2}.$$

TABLE OF SQUARES AND SQUARE ROOTS*

x	x^2	$\sqrt{x}$	x	x^2	$\sqrt{x}$
1	1	1.000	26	676	5.099
2	4	1.414	27	729	5.196
3	9	1.732	28	784	5.292
4	16	2.000	29	841	5.385
5	25	2.236	30	900	5.477
6	36	2.449	31	961	5.568
7	49	2.646	32	1,024	5.657
8	64	2.828	33	1,089	5.745
9	81	3.000	34	1,156	5.831
10	100	3.162	35	1,225	5.916
11	121	3.317	36	1,296	6.000
12	144	3.464	37	1,369	6.083
13	169	3.606	38	1,444	6.164
14	196	3.742	39	1,521	6.245
15	225	3.875	40	1,600	6.325
16	256	4.000	41	1,681	6.403
17	289	4.123	42	1,764	6.481
18	324	4.243	43	1,849	6.557
19	361	4.359	44	1,936	6.633
20	400	4.472	45	2,025	6.708
21	441	4.583	46	2,116	6.782
22	484	4.690	47	2,209	6.856
23	529	4.796	48	2,304	6.928
24	576	4.899	49	2,401	7.000
25	625	5.000	50	2,500	7.071

*Square roots rounded to nearest .001.

x	x^2	$\sqrt{x}$	x	x^2	$\sqrt{x}$
51	2,601	7.141	76	5,776	8.718
52	2,704	7.211	77	5,929	8.775
53	2,809	7.280	78	6,084	8.832
54	2,916	7.348	79	6,241	8.888
55	3,025	7.416	80	6,400	8.944
56	3,136	7.483	81	6,561	9.000
57	3,249	7.550	82	6,724	9.055
58	3,364	7.616	83	6,889	9.110
59	3,481	7.681	84	7,056	9.165
60	3,600	7.746	85	7,225	9.220
61	3,721	7.810	86	7,396	9.274
62	3,844	7.874	87	7,569	9.327
63	3,969	7.937	88	7,744	9.381
64	4,096	8.000	89	7,921	9.434
65	4,225	8.062	90	8,100	9.487
66	4,356	8.124	91	8,281	9.539
67	4,489	8.185	92	8,464	9.592
68	4,624	8.246	93	8,649	9.644
69	4,761	8.307	94	8,836	9.695
70	4,900	8.367	95	9,025	9.747
71	5,041	8.426	96	9,216	9.798
72	5,184	8.485	97	9,409	9.849
73	5,329	8.544	98	9,604	9.899
74	5,476	8.602	99	9,801	9.950
75	5,625	8.660	100	10,000	10.000

SOURCES

The following have served as sources for one or more problems.
Page numbers indicate the place(s) in this book where the problems
can be found.

Auerbach, Alvin B. and Vivian Shaw Groza. Introductory Mathematics
   for Technicians. New York: Macmillan, 1972. (p. 509, 667, 687)

Batschelet, Edward. Introduction to Mathematics for Life Scientists.
   New York: Springer-Verlag, 1973. (p. 666)

Chatterjee, S. "Estimating the Size of Wildlife Populations," in
   Statistics by Example, Frederick Mosteller et al., editors.
   Reading, MA: Addison-Wesley, 1973. (p. 260, 261)

Luy, Jack. Practical Problems in Mathematics for Carpenters.
   Albany, NY: Delmar Publications, 1973. (p. 192)

Mosteller, Frederick, Robert E. K. Rourke, and George B. Thomas,
   Jr. Probability with Statistical Applications, second edition.
   Reading, MA: Addison-Wesley, 1970. (p. 447)

The following have served as sources of data.

Asimov, Isaac. The Realm of Measure. New York: Fawcett World
   Library, 1967.

Bell, Max. Mathematical Uses and Models in Our Everyday World.
   SMSG Studies in Mathematics, Vol. XX. Stanford, CA: SMSG,
   1973.

Chrystal, G. Algebra. Parts I and II. Seventh Edition. New York:
   Chelsea, 1964.

Engineering Concepts Curriculum Project. The Man-Made World.
   New York: McGraw-Hill Publishing Co., 1971.

Fargis, Paul, ed. The Consumer's Handbook. New York: Hawthorne
   Books, 1974.

The Columbia-Viking Desk Encyclopedia. New York: Viking Press, 1953.

The World Almanac and Book of Facts for 1966. Luman H. Long, editor. New York: New York World-Telegram, 1966.

The World Almanac and Book of Facts, 1973 edition. New York: Newspaper Enterprise Association, 1972.

The Official Associated Press Almanac 1975. Maplewood, NJ: Hammond Almanac, Inc., 1974.

The Guiness Book of World Records, 10th edition. Norris and Ross McWhirter, editors. New York: Bantam Books, 1971.

Official Major League Baseball Record Book, 1972 edition. New York: Fawcett World Library, 1972.

Better Homes and Gardens Cookbook

Changing Times, September, 1970.

Schwartz, Harry. "America the Mathematized." The New York Times, November 7, 1969.

The Chicago Sun-Times, August 25, 1974.

Commonwealth Edison rates pamphlet, April, 1974.

Holiday Inn directory, February 1 - May 31, 1973.

Telebriefs. Vol. 37, No. 9. Chicago: Illinois Bell Telephone Co., October, 1975.

747

INDEX OF SYMBOLS

ACKNOWLEDGMENTS

In 1969, when I came to the University of Chicago, I was in the process of finishing some work in geometry. That work contained few applications even though the concepts in the course lend themselves to applications. But at the time no one had ever convinced me that work with applications could be as beneficial as the work "without" applications found in courses of that time.

By 1971, when I was doing some work in second-year algebra, my colleague Max Bell had convinced me that the standard "word problems" (these being the only ostensible connection between algebra and the real world) were not applications at all and may do more harm than good by distorting the relationships which exist between mathematics and the real world. It was relatively easy to modify the word problems in that course to make them "more real," but it was also clear that the students in that course are a brighter minority of mathematics students and something more would be needed for the majority.

Looking at the real world and the increased amount of data citizens are expected to process (or else be bewildered by) has convinced us (and many others) that work with statistics is essential for all students. Talks with statisticians, furthermore, convinced me that the statistics _they_ wanted to see in schools was much simpler than that found in statistics courses at the university level. Much of the statistics they talked about could easily be done by first-year algebra students; some seemed reasonable even for less mathematically mature students.

Thus by 1972, the seed had been planted for this course - a first-year algebra course in which applications would be given heavy emphasis and containing some statistics (with the corresponding probability). But material cannot be added except at the expense of other material. It was easy to decide what would not be given heavy weight in the course. The most difficult material turned out to be exactly that material which is least needed in later courses or in applications. (Perhaps the connection is more causal than casual; tedious mathematics usually has the fewest applications or is now done by computers.)

Preparation for writing this course began in 1973 with courses and workshops at the University of Chicago. National Science Foundation support has been present from the first writing in 1974 to the present. This course might have been written without that help, but certainly would not have been finished so quickly and probably would have greatly suffered in quality. My gratitude for their support extends warmly to Lauren Woodby of Michigan State University, James Wilson of the University of Georgia, and Joseph Payne of the University of Michigan.

Parts or all of the manuscript in its various drafts have been read by a number of university people. Particular thanks must go to

Max Bell, whose earlier work in uses of numbers written for SMSG turned out to be the single most influential idea in the development of this book. Others of great help have included William Kruskal and Pamela Ames of the University of Chicago, James Schultz of Ohio State University, and Jane Swafford of Northern Michigan University.

Pilot drafts of this manuscript were taught in schools by Derk Prentiss of Addison Trail H.S., Addison, IL; Lynn Lazuka Banasik of Unity H.S., Chicago; and Marie Hill and Jewel Crosby of the Proviso Twp. High Schools in Maywood and Hillside, IL. These teachers seldom received material more than a few days in advance and deserve special plaudits and thanks for their willingness to participate and their countless suggestions for improving this manuscript.

The project itself has been fortunate to have had the services of John Easton, Research Assistant par excellence, and Dana Gregorich, who has typed both pilot drafts and a multitude of other project-related materials. The present manuscript was typed by Mary McKillip and Linda Collinwood. All of these people have had to work under the constant pressure of deadlines; their help is very much appreciated.

But, as must be the case, the responsibility for this entire preliminary draft lies with its author, and all comments, suggestions, or criticism should be directed to me.

Zalman Usiskin
University of Chicago
October, 1976